DINOSAURUS

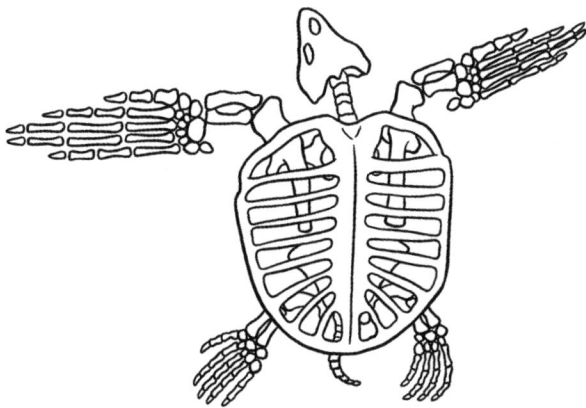

Smithsonian
Books

© 2025 by UniPress Books Ltd

Published in the United States and Canada by Smithsonian Books
PO Box 37012, MRC 513
Washington, DC 20013
smithsonianbooks.com

Director: Carolyn Gleason
Production Editor: Julie Huggins
Project Editor: Richard Webb
Designed by Clare M. Barber

This book may be purchased for educational, business,
or sales promotional use. For information please write the
Special Markets Department at the address or website above.

Library of Congress Cataloging-in-Publication Data available upon request.

Hardcover ISBN: 978-1-58834-798-5

Printed in China, not at government expense
29 28 27 26 25 1 2 3 4 5

MIX
Paper | Supporting
responsible forestry
FSC
www.fsc.org   FSC™ C005748

DINOSAURUS

A PREHISTORIC DICTIONARY

RHYS CHARLES

ILLUSTRATIONS BY SAVANNAH STORM

Smithsonian Books
Washington, DC

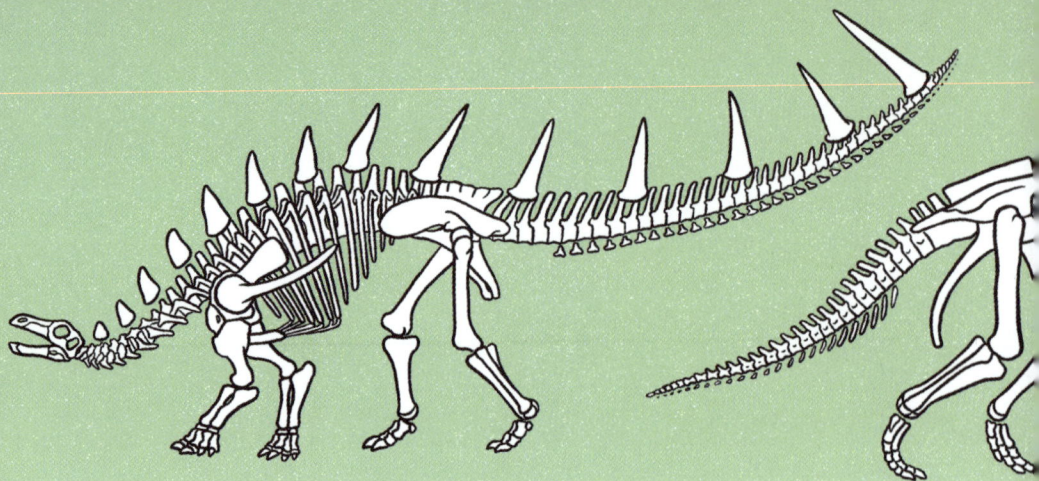

CONTENTS

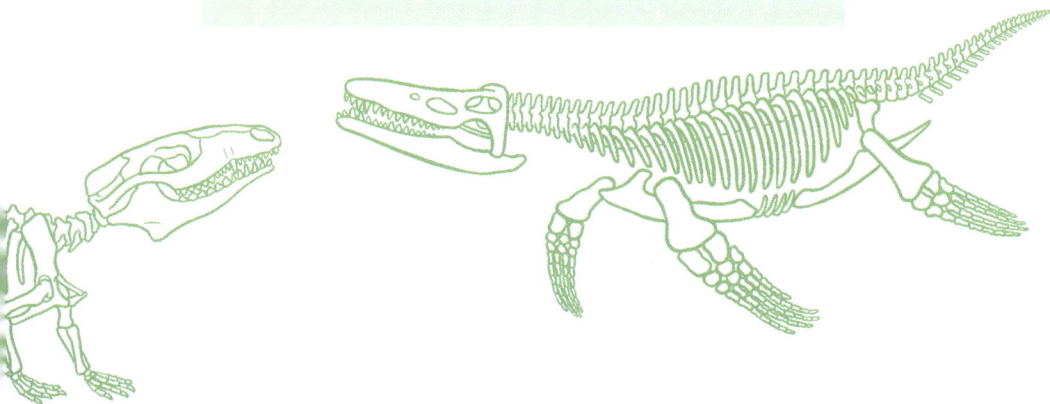

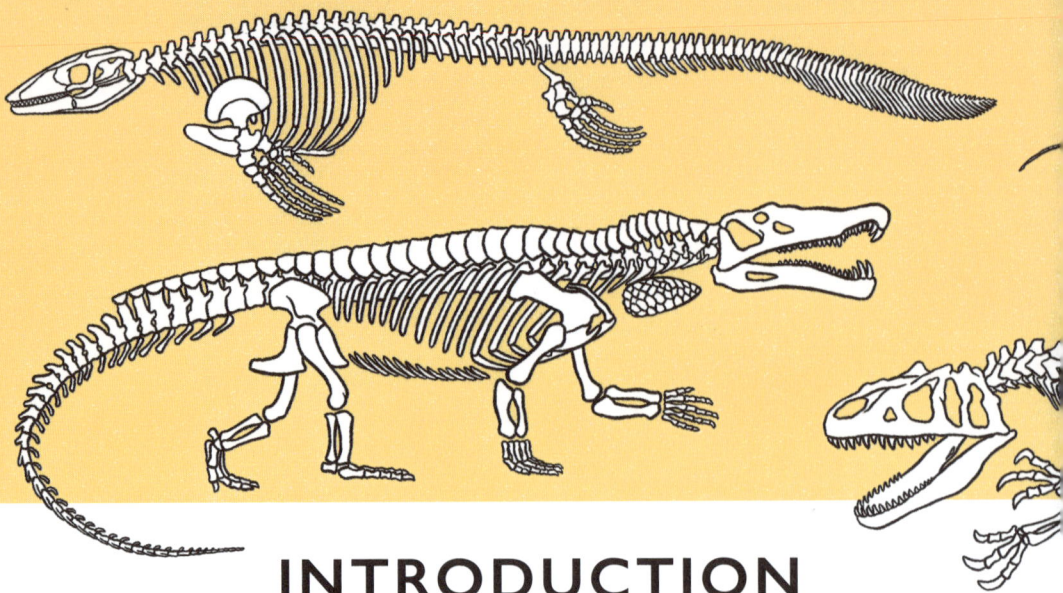

INTRODUCTION

Ever since the very first fossils were uncovered, people in every corner of the globe have been fascinated by dinosaurs.

The long-extinct creatures have seemingly invaded every aspect of our culture, from film and media to toys and textiles. But with this immense popularity can also come confusion: It's a challenge to separate the fact from the fiction, the real animals from the almost mythological beings they have become.

The world the dinosaurs knew was certainly very different from our own and, over the past two hundred years, scientific research has revealed a staggering amount about it. Studies of the evidence the dinosaurs left behind has allowed paleontologists to breathe life back into these prehistoric beasts.

Experts have been able to reconstruct how they looked, placing not only the muscle, but fat, skin, and feathers onto the bone. The relationships between dinosaur species have been reconstructed to show their place in the great tree of life, between their reptilian ancestors and their bird descendants. We've even been able to examine clues in the rock that reveal glimpses into their behaviors and social interactions.

However, this amazing information can seem impenetrable at times due to the number of scientific and technical terms with which it frequently comes hand in hand. The sheer amount of jargon can sometimes act as a barrier to anyone who wants to casually visit the world of the dinosaurs, instead of setting up camp for years to get a doctorate in the field.

The study of prehistoric life is so vast, and covers so many aspects of science, that a whole new dictionary has effectively been written on the subject to help researchers navigate the science and communicate their theories. This book aims to break down some of those names and terms to explore the meaning behind them, and make the dinosaur age more accessible.

In the first chapter, "Dinosaurs," we take a look at some of the more iconic members of the dinosaur family, revealing the animals behind the household names and unmistakable silhouettes. Along the way are many lesser-known names which are every bit as fascinating, and whose stories help fill in key details to understand the bigger picture of the dinosaurs.

Despite what some might think, however, it wasn't exclusively dinosaurs that were roaming the Earth during prehistoric times.

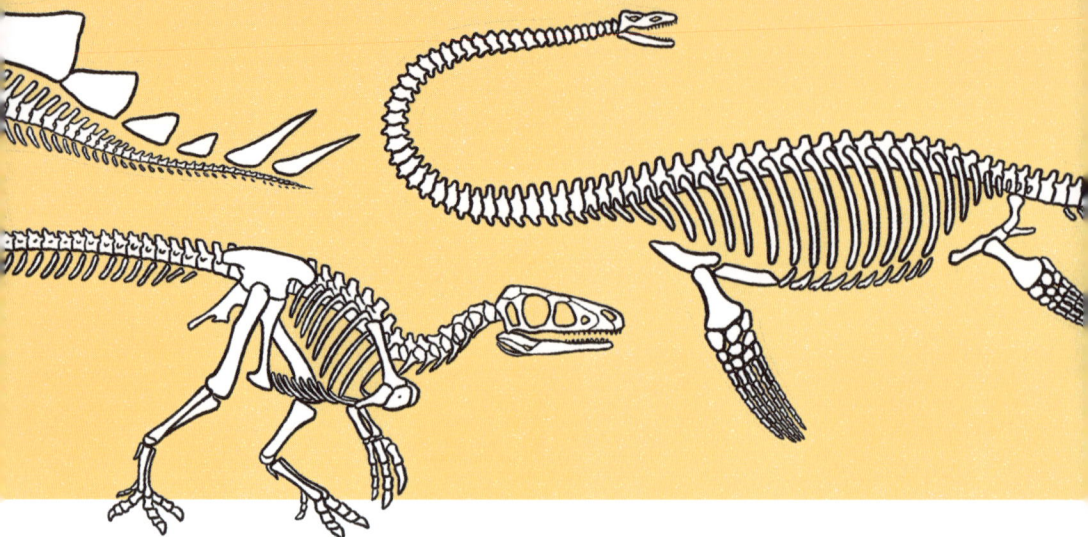

In Chapter 2, "Prehistoric Beasts," we uncover some of the amazing animals that shared the planet with them, on the land, in the sea, and in the skies above, coming together to form one united prehistoric ecosystem. Some of the animals described here will seem familiar, not too removed from what you might encounter living in the wilds today. Others, though, are completely unique, with any relations having long since disappeared from existence. A few of them are even regularly mistaken for dinosaurs themselves, despite being almost totally unrelated and far apart in evolutionary terms.

Any conversation about dinosaurs is invariably going to involve topics that span several millions of years, as they existed over such a large span of time and so long ago compared to humans. In Chapter 3, "Events and Time," we break down this concept of "deep time" into manageable chunks, explaining the major events and the names scientists give to particular windows of prehistory. In sequence, these events tell the story of the dinosaurs' rise to dominance, their fall from grace, and how this would eventually pave the way for our own story.

Just as it is today, it is useful to have a map complete with place names and landmarks in order to understand the prehistoric world. However, due to the ever-changing nature of our planet, the map of the dinosaur age would seem totally alien to us, the drifting continents and shifting seas creating and destroying whole regions over the millennia. In Chapter 4, "Prehistoric Planet," we delve into some of the features of this prehistoric map, and chart onto it the locations where some of the best dinosaur fossils can be discovered today.

Finally, Chapter 5, "Scientific Terms," is dedicated to the latest discoveries in the modern and fast-growing science of paleontology, explaining the meaning behind some of the most commonly used vocabulary. The techniques, types of evidence, and terms defined here are what have helped us to understand the dinosaurs to a level of detail that would have seemed impossible even just a decade ago. Here lies much of the science nitty-gritty, the foundations that underlie our new thinking of this old world.

When put together, the words collected here build up a rich history of life on Earth, and the magnificent array of dinosaurs that once roamed it.

TIMELINE

PHANER

PALEOZOIC

CAMBRIAN

ORDOVICIAN

SILURIAN

DEVONIAN

CARBONIFEROUS

PERMIAN

538.8 million years ago (mya) — Cambrian Explosion

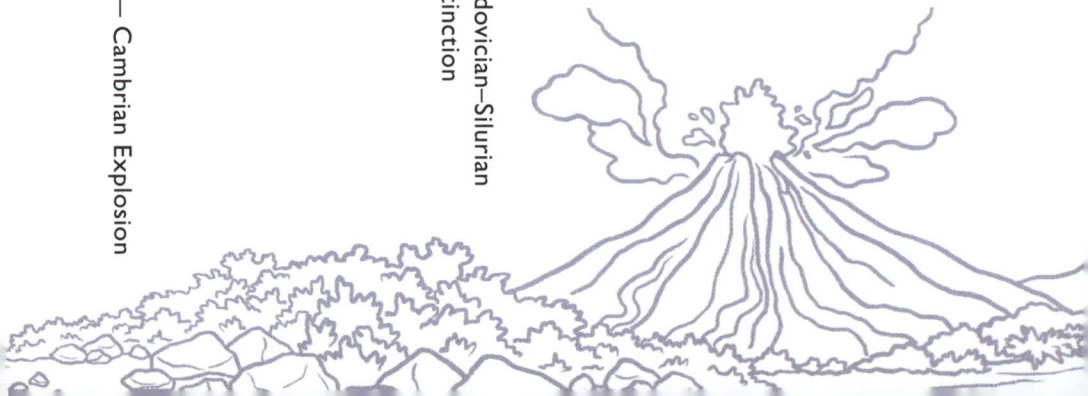

485.4 mya

443.8 mya — Ordovician–Silurian Extinction

419.2 mya

358.9 mya

298.9 mya

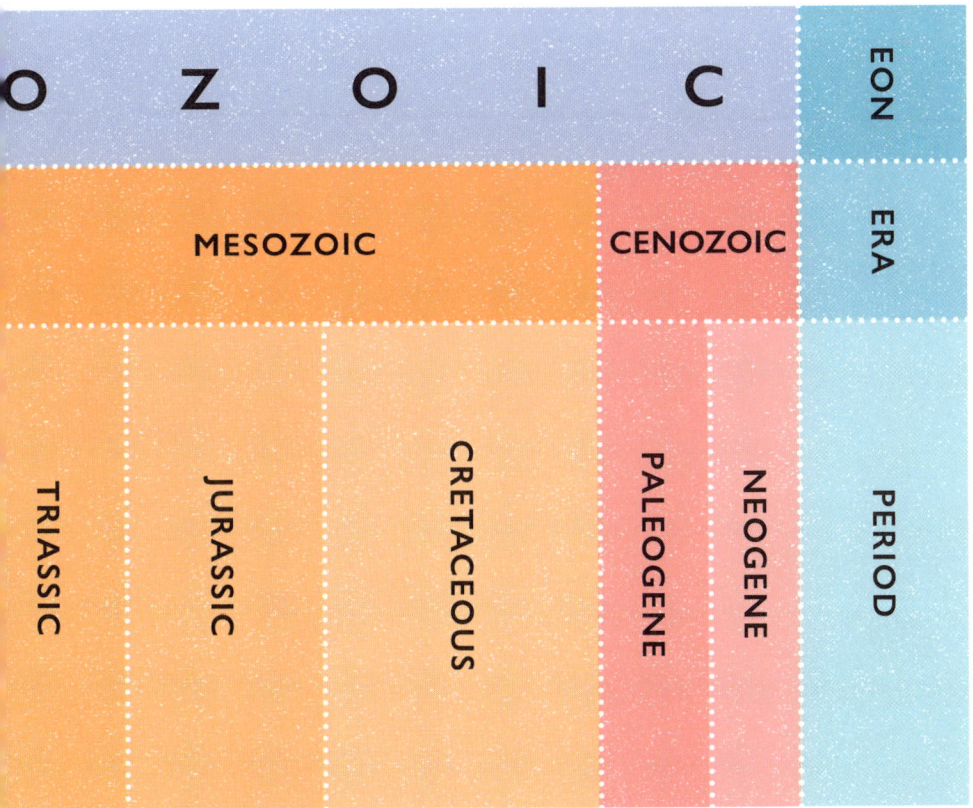

O Z O I C

EON	ERA	PERIOD
	MESOZOIC	TRIASSIC
	MESOZOIC	JURASSIC
	MESOZOIC	CRETACEOUS
	CENOZOIC	PALEOGENE
	CENOZOIC	NEOGENE

Permian–Triassic

201.4 mya —
Triassic–Jurassic
Extinction

145.0 mya

Angiosperm Revolution

66.0 mya —
Cretaceous–Paleogene
Extinction

23.0 mya

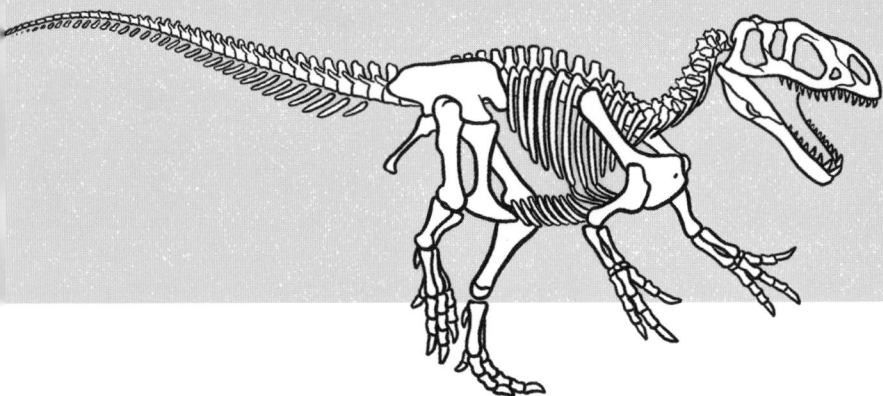

CHAPTER 1
DINOSAURS

INTRODUCTION

For more than 160 million years, during the Mesozoic era (252–66 million years ago), dinosaurs were the dominant vertebrate life on Earth.

The dinosaurs evolved into a host of different shapes and sizes, and some of them became the largest animals ever to have walked the Earth. Others evolved a breakthrough adaptation that would completely change how they were able to regulate their body temperature and display to each other, and eventually enabled them to take to the skies. This adaptation was the feather, a feature unique to the dinosaur group (which technically includes all living and extinct species of birds).

As a whole, the dinosaurs shared certain characteristics that set them apart from the rest of the reptiles. Chief among these was the shape of their thighbone (femur), which allowed their legs to be held directly underneath their bodies—as opposed to sprawling out to the sides, like in lizards and crocodiles.

Paleontologists continue to adjust the dinosaur family tree, identifying differences between individuals that further split the dinosaurs into distinct groups. Some of the differences can be prominent anatomical structures, such as massive bone plates erupting from the body. But equally, they can be minute, measured in the subtle differences in the shape and positioning of certain bones.

At the time of writing, just over 1,000 species of dinosaur have been named and described, and in this chapter we will highlight sixty of the most intriguing genera, stretching from their first appearance 230 million years

ago in the Triassic period, through the Jurassic, and up to the extinction event which wiped them out at the end of the Cretaceous, 66 million years ago.

CLASSIFICATION

Theropods, sauropodomorphs, and ornithischians are three families of dinosaurs that each include a variety of dinosaur species. The theropods were generally meat-eaters, known for mouths full of sharp teeth, standing on two legs, and with hands regularly tipped with pointed claws. This group includes such famous predators as the huge *Tyrannosaurus rex* and agile *Velociraptor*, as well as a few other surprising additions.

Sauropodomorphs were large plant-eaters with long necks for browsing foliage. These dinosaurs reached truly enormous sizes, with their immense bodies supported by four columnar legs.

On the other side of the family tree were the ornithischians, a group that shared a birdlike arrangement of hip bones. Predominantly herbivores, this superfamily contains the frilled and horned ceratopsians, dome-headed pachycephalosaurs, armored ankylosaurs, duck-billed hadrosaurs, and the highly ornamented stegosaurs.

Together, these dinosaur lineages formed one of the most diverse, successful, and awe-inspiring groups of animals to have ever existed.

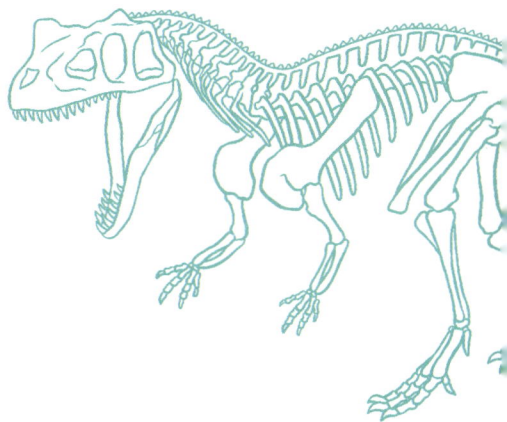

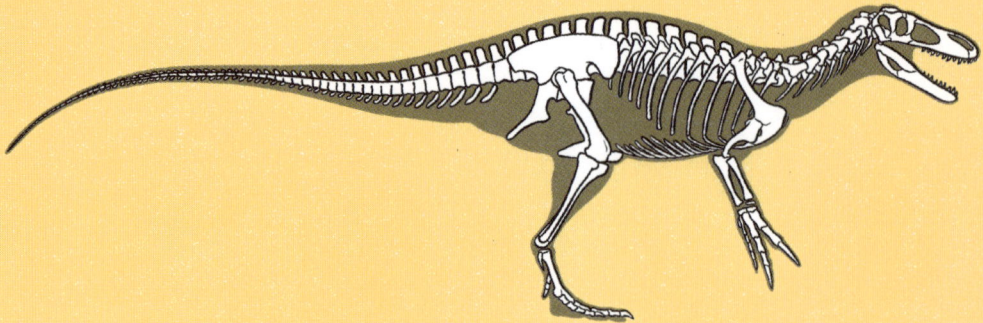

AEROSTEON

AIR-OS-TEE-ON

FIRST DISCOVERED: 1996 **RANGE:** ARGENTINA

The connection between dinosaurs and birds may now be well known, but some shared features can still surprise us. The predator *Aerosteon* is a great example of this, and a specimen uncovered in South America gives us further proof of the evolutionary relationship between the two species.

Evidence for dinosaur breathing systems is extraordinarily rare, as soft tissues (like lungs) almost always decay long before they get the chance to fossilize. As a result, researchers instead take clues from the structure of the bones, and hollow spaces, like air pockets discovered along the wishbone and the top of the spine, suggest *Aerosteon* drew breath in a similar way to modern birds. While humans and other mammals breathe in a tidal, "in and out" fashion, avian breathing uses a system of bellows, or air sacs, that involves a unidirectional flow of air, meaning their lungs don't have to expand. It's much more efficient and it's why birds can fly higher than, say, bats.

Paleontologists also discovered air sacs that came around the edge of *Aerosteon*'s body and into the belly ribs, creating a system of air tubes under the skin. It is thought this aided temperature regulation, which, given *Aerosteon* weighed as much as an elephant and had feathers and no sweat glands, would have come in handy. Indeed, if *Aerosteon* and other dinosaurs did have this avian breathing, it would have facilitated an active lifestyle, likely making this 25-foot-long (7.6 m) dinosaur a fast and dangerous predator.

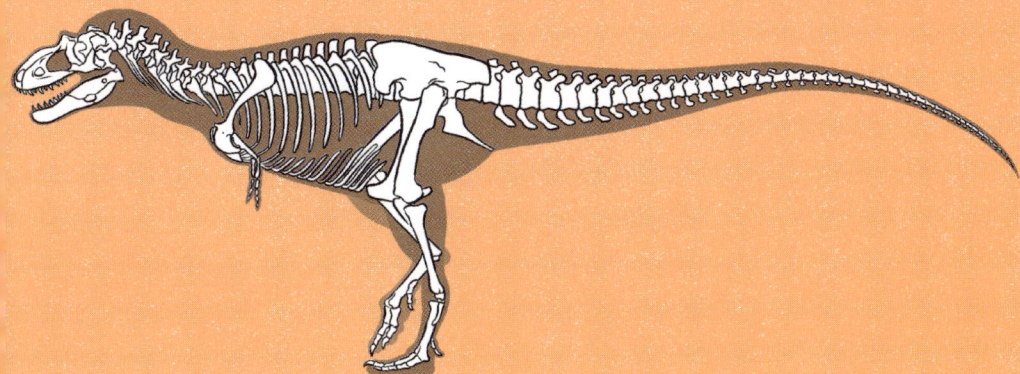

ALBERTOSAURUS

AL-BER-TOE-SAW-RUS

FIRST DISCOVERED: 1884 **RANGE:** CANADA AND USA

Named for the Canadian province Alberta, *Albertosaurus* was a top predator in the wilds of what is now Canada 70 million years ago. Part of the Tyrannosauroidae family, it shares many features with its cousin to the south, *Tyrannosaurus rex*, including the famously short arms.

Although lighter and leaner than a *T. rex*, its teeth show a characteristic "D" cross-sectional shape and their skulls are built strong and resistant to bending pressures—adaptations that make them perfectly evolved for grasping and holding onto struggling prey, leaving little doubt that *Albertosaurus* was an active hunter of other large dinosaurs.

Fossils of this dinosaur can be found across a huge geographical range (estimated at 96,500 square miles, or 250,000 sq km), which shows how successful it was. In the Horseshoe Canyon Formation of Edmonton, where changes in climate drove away many other species of dinosaur, *Albertosaurus* stood strong, able to thrive in multiple different habitats.

Bite marks recently discovered on the hip bones of *Albertosaurus* remains give a glimpse into the brutal lives of predatory dinosaurs. There is speculation that there may have even been some dinosaur-on-dinosaur hunting, since these marks were of the same dimensions as *Albertosaurus*'s bite.

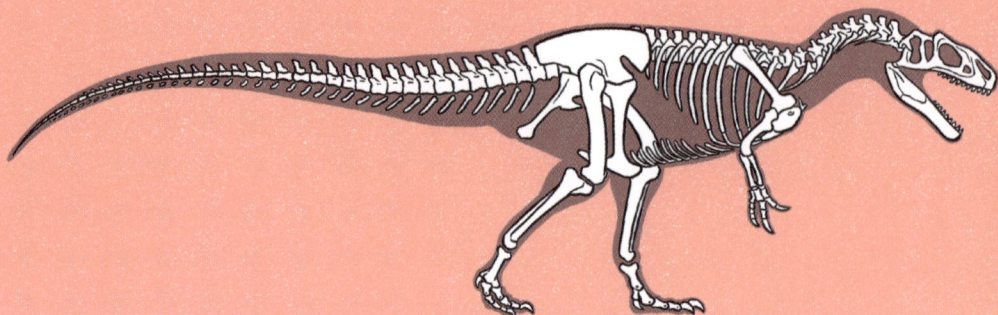

ALLOSAURUS

ALL-O-SAW-RUS

FIRST DISCOVERED: 1877 **RANGE:** USA

Often considered the top predator of the Jurassic period, *Allosaurus* was a formidable dinosaur in ancient North America, and is the most commonly found theropod of the famed Morrison Formation—the area where most dinosaur fossils have been discovered in the US.

Extensions above the eyes of *Allosaurus*, covered in the protein keratin (like claws and our fingernails), gave it an almost horned appearance. As the horns were likely too small to be used in combat, they were probably display structures.

Thanks to the incredible abundance of *Allosaurus* fossil remains, scientists have been able to study the rapid growth rates of this dinosaur, as well as the injuries many of them picked up in life. Several show evidence of multiple broken bones from particularly traumatic events, possibly fights with prey or even other allosaurs. One unfortunate individual seems to have had spinal arthritis.

Some fossil bones of giant long-necked sauropod dinosaurs of the Morrison Formation show evidence of predation by *Allosaurus*, with tooth marks etched onto the bone. The risk of severe injury or death, it seems, wasn't enough to put off *Allosaurus* from tackling the toughest of prey.

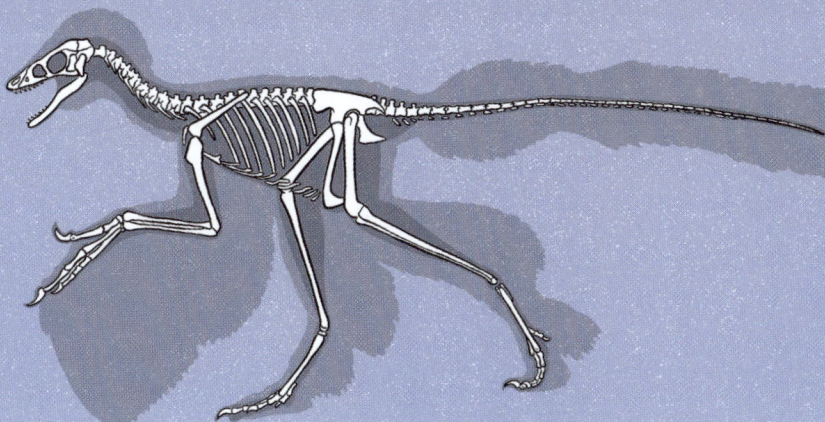

ANCHIORNIS

AN-KEE-OR-NUS

FIRST DISCOVERED: 2009 **RANGE:** EASTERN CHINA

Covered in feathers, at first glance *Anchiornis* appears much more bird than dinosaur, and in this way beautifully illustrates how blurry the line between the two groups is. Whereas birds only have wings on their front limbs, however, *Anchiornis* has them on its hind legs too, making it one of several "four-winged dinosaurs."

Anchiornis, or *Anchiornis huxleyi*, to give the species its full name, was denominated in honor of Thomas Henry Huxley, a nineteenth-century biologist who was one of the first champions of Darwin's theory of natural selection. It was also Huxley who, in 1868, became the first to suggest that birds were in fact types of dinosaur, over a hundred years before that theory was truly accepted.

One of the most fascinating things about *Anchiornis* is that, thanks to the exceptional preservation of its feathers and the pigmentary structures within

them, paleontologists have been able to reconstruct its color. *Anchiornis* was predominantly black and white, with patches of red, likely used for display, around the head.

This remarkable preservation means scientists can reconstruct this species of dinosaur more accurately than almost any other.

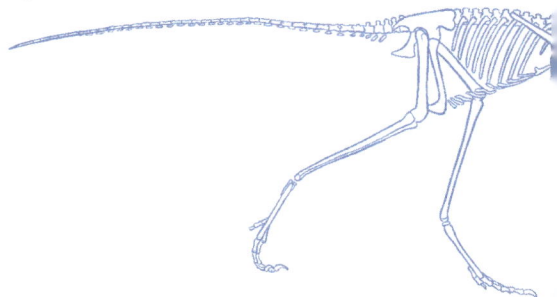

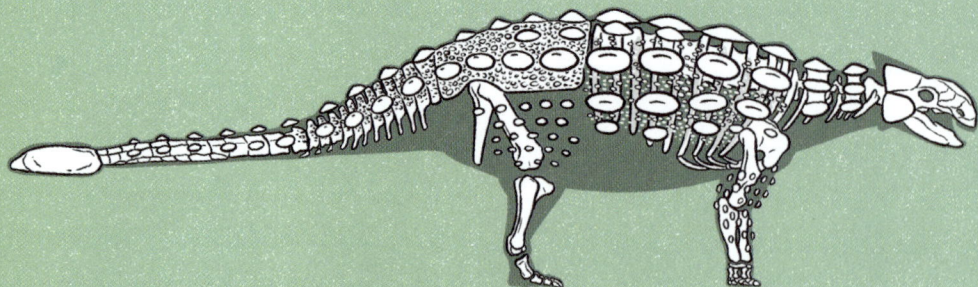

ANKYLOSAURUS

AN-KEE-LO-SAW-RUS

FIRST DISCOVERED: 1906 **RANGE:** CANADA AND USA

Dinosaurs are well known for their diverse set of weaponry and defensive structures, and no species took this to a greater extreme than *Ankylosaurus*.

This tank of the late Cretaceous was covered in thick plates of armor called osteoderms, and had a tail tipped off by an impressive bone club. The osteoderms of *Ankylosaurus* are relatively smooth compared to those of related genera, a handy tip for telling ankylosaurs apart.

The obvious thought on seeing this club is that it would have been used as some kind of defensive weapon, swinging from side to side to deter attackers of any size. This is a surprisingly difficult thing to prove empirically from fossils alone, but scientists still mostly agree that this was indeed one of its uses.

Another theory is that it could have been used in competition within its own species. Some ankylosaurs show evidence of damage to their armor concentrated around the hips, suggesting that the clubs were being used in a kind of ritualistic combat. If it wasn't as organized, you would expect to see injuries spread across the body, rather than focused in one place like this.

Due to its patchy fossil record, only one tail club confirmed to have come from *Ankylosaurus* itself is known, despite being its most famous feature. Discovered in 1910, just four years after the first fossils of the genus were uncovered, no other club has been found in the more than one hundred years since, and evidence from other related species has been used to help inform today's reconstructions.

Unlike some of their smaller (and potentially less well defended) relatives, it's thought that *Ankylosaurus* lived a solitary life in adulthood, as opposed to travelling in herds. At up to over 20 feet (6 m) in length and weighing 7.9 tons (8,000 kg), however, it's fair to say that *Ankylosaurus* could have handled itself pretty well.

Living at the very end of the Cretaceous period in what is now North America, *Ankylosaurus* was part of the most famous community of dinosaurs, alongside *Tyrannosaurus* and *Triceratops*. The largest and last of the ankylosaurid family, *Ankylosaurus* lived on the planet until the very day the asteroid struck.

SELF-DEFENSE

Composed of dense bone and thick cartilage, *Ankylosaurus*'s clubbed tail was a potent weapon, delivering crushing strikes that would have quickly disabled most threats. In combination with its numerous osteoderms, *Ankylosaurus* presented a formidable opponent to any predator foolhardy enough to attack.

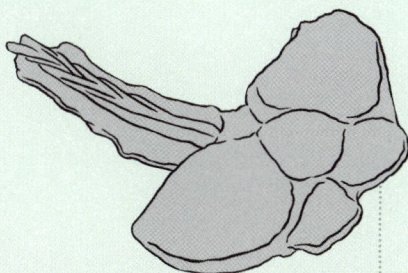

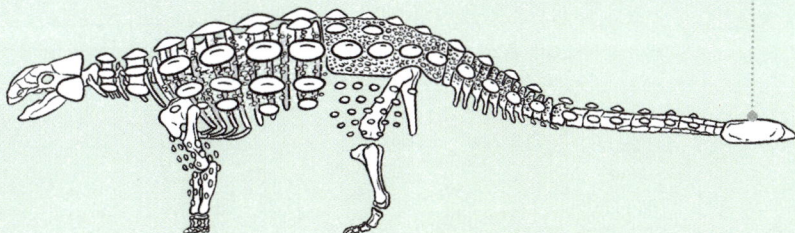

ARGENTINOSAURUS

AR-GEN-TI-NO-SAW-RUS

FIRST DISCOVERED: 1987 **RANGE:** ARGENTINA

Declaring definitively which dinosaur was the largest ever to have existed is naturally a difficult and controversial process, but one of the names that consistently appears in the running, weighing in at around 93 tons (94,492 kg), is undoubtedly *Argentinosaurus*.

A single femur from this giant can measure 8.2 feet (2.5 m) in length, and estimates of total body size range all the way to 115 feet (35 m). If correct, this would make *Argentinosaurus* potentially the largest animal to have ever walked the planet. However, with so much of this based on fragmentary fossil evidence, we can't be certain of accuracy.

Fossils of this Cretaceous giant were first uncovered in Argentina in the 1980s and assigned to a group fittingly called the titanosaurs. Their small heads lacked the musculature for complex chewing, and instead they relied on their huge guts to break down and extract the maximum energy from their plant food, allowing them to grow to an epic scale. Their long necks also allowed for efficient banquet feeding, essentially consuming as much food as possible while standing in one place.

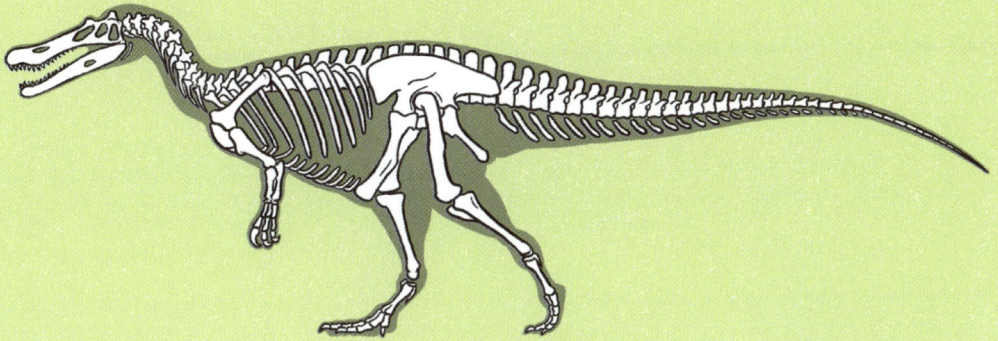

BARYONYX

BA-REE-O-NYX

FIRST DISCOVERED: 1983 **RANGE:** UK AND WESTERN EUROPE

One of the largest theropods found in Britain, the initial discovery of a single huge curved claw caused a sensation when it was unveiled in 1986. Over a foot (30 cm) long, the purpose of this terrifying weapon was revealed when the skull material was found.

The distinctly crocodile-like face of *Baryonyx* gives its family and feeding choices away. *Baryonyx* was a spinosaur, and lived largely off a diet of fish, using that heavy claw to hook prey out of the waterways. In one specimen, the bones of a juvenile iguanodontid dinosaur were found within the rib cage of the *Baryonyx*. Up to 33 feet (10 m) from nose to tail and armed with those formidable weapons, there's no reason to say that other dinosaurs wouldn't occasionally have featured on the menu of this early Cretaceous hunter.

There is some evidence that *Baryonyx* had a heightened ridge along its back, but certainly nothing like the full sails seen in some of its relatives.

In 2021, the announcement of two more potential spinosaurs from the Isle of Wight, UK, suggested that *Baryonyx* didn't have the waterways of what is now Northern Europe all to itself.

BRACHIOSAURUS

BRACH-I-O-SAW-RUS

FIRST DISCOVERED: 1900 **RANGE:** USA

Perfectly evolved to reach even the highest branches, the Jurassic giant *Brachiosaurus* was the size of two giraffes stacked one on top of the other. Its name means "Arm Lizard," in reference to its gigantic front legs, which were significantly longer than those at the back.

Study of the microscopic wear patterns on its peg-like teeth shows that *Brachiosaurus* was perfectly evolved for raking the leaves from trees, exploiting a niche in the environment not available to most other herbivorous dinosaurs.

Brachiosaurus and its relatives were once fancifully hypothesized to have been too big to walk on land, and were said to have instead lived in the water, using their 28-foot (8.5-m) necks like snorkels. It didn't take much research to rule this out as an impossibility, even if the myth persists.

Clues as to how these giants survived on land can be found in their bones. The ribs of *Brachiosaurus* show a highly pneumatized, sponge-like structure, making them far lighter than might be expected. It's a similar system utilized by birds, and a surprising connection between a 1-oz (30-g) sparrow and a 35-ton (35,561-kg) dinosaur.

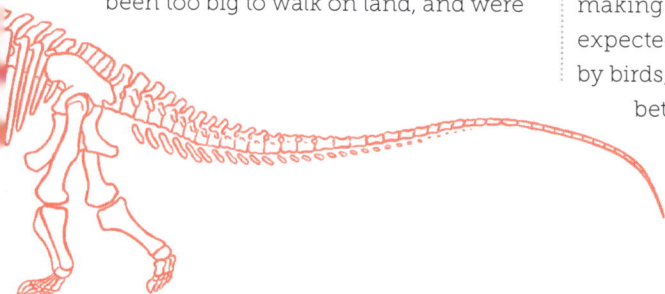

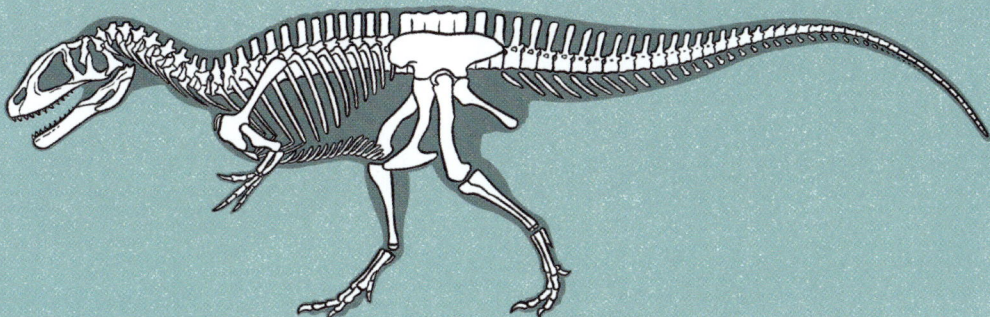

CARCHARODONTOSAURUS

CAR-KA-RO-DON-TOE-SAW-RUS

FIRST DISCOVERED: 1924 **RANGE:** NORTH AFRICA

A mouthful of shark-like teeth give this enormous predator its name. Measuring up to 40 feet (12 m) from snout to tail, *Carcharodontosaurus* is one of only a handful of carnivorous dinosaurs estimated to have grown larger than *T. rex*. A top predator in what is now North Africa, one species has been named for the desert across which it can be found: *C. saharicus*.

Sadly, the original material of this dinosaur, first discovered in 1924, was lost during bombing in World War II. Thankfully more fossils have been unearthed in the years since, allowing the reconstruction of this formidable dinosaur, to the extent that paleontologists have been able to calculate its jaw and neck strength. *Carcharodontosaurus* was able to lift over 880 lbs (400 kg) with its jaws. That's about the same weight as a polar bear, one of the top predators in the world today.

A thick ridge of bone along the top of the skull has raised the possibility that this dinosaur might even have competed through headbutting. For this dinosaur, it was probably not a bad idea in fights when the alternative was those deadly jaws.

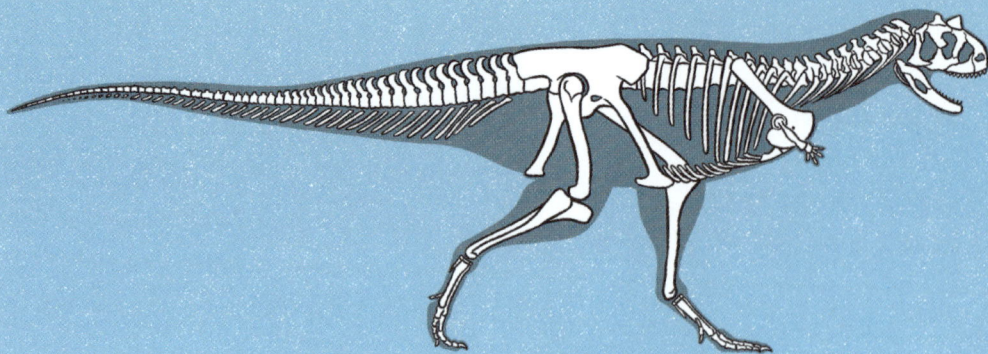

CARNOTAURUS
CAR-NO-TAW-RUS
FIRST DISCOVERED: 1984 **RANGE:** ARGENTINA

With a name meaning the "Carnivorous Bull" and first reported as having "Devil Horns," this Cretaceous predator from what is now South America is regularly cast as a villain in dinosaur pop culture.

Carnotaurus is the most famous member of a group of dinosaurs called the Abelisauridae, all of which were predators with thickened skull regions, though no others would flaunt those iconic horns. However, bone structure analysis suggests it only had a small sheath covering, rather than a full, cow-like long horn. Ornamentation went beyond just the head too, as the body of the animal was covered in randomly distributed special bony scales, likely for defense.

The arms of Carnotaurus were tiny, comparatively smaller even than those of T. rex, but it made up for this with powerful legs, allowing it to chase its prey and hunt at speed.

Carnotaurus and its abelisaurid relatives were the top predators of ancient South America for the latter half of the Cretaceous period, taking over from the giant carcharodontosaurs and staying at the top of the food chain until the extinction of all the dinosaurs 66 million years ago.

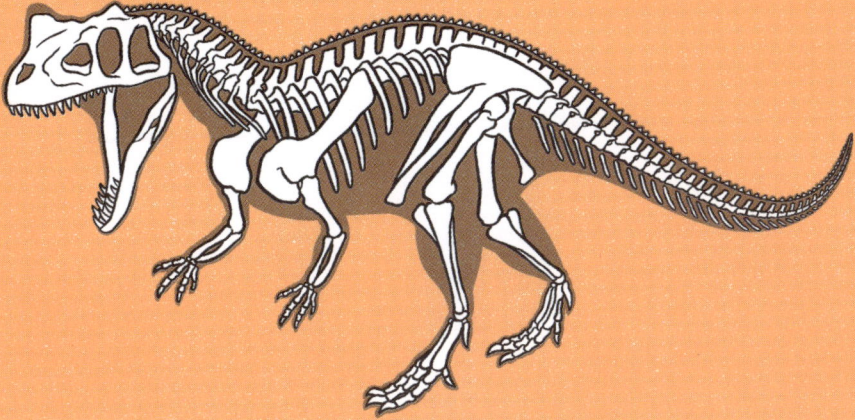

CERATOSAURUS

SAIR-A-TOE-SAW-RUS

FIRST DISCOVERED: 1883 **RANGE:** USA

The bulbous ridge of bone atop its nose gives *Ceratosaurus* a fairly unique skull profile among theropods. Once thought to be the base for a large horn the dinosaur could use for attacking prey, it's now thought it much more likely served as a display structure, as opposed to the species being some kind of reptilian unicorn.

In addition to its horn, two smaller bone ridges appear directly above the eyes. Even if only small protrusions, all three show traces of blood vessels, indicating they were covered in a smooth sheath of the protein keratin, like claws.

The rest of *Ceratosaurus*'s body has fueled some interesting hypotheses about how it may have hunted. For one, *Ceratosaurus* had a broader and less rigid tail than many of the other carnivorous dinosaurs, which has caused some paleontologists to speculate that it may have been a shallow-water specialist, maybe even propelling itself through water, almost like a crocodile.

The verdict is still out on this aquatic theory. However, if true, it would certainly have helped *Ceratosaurus* avoid competition for food with the larger *Allosaurus*, with whom it shared the hunting grounds of Jurassic North America.

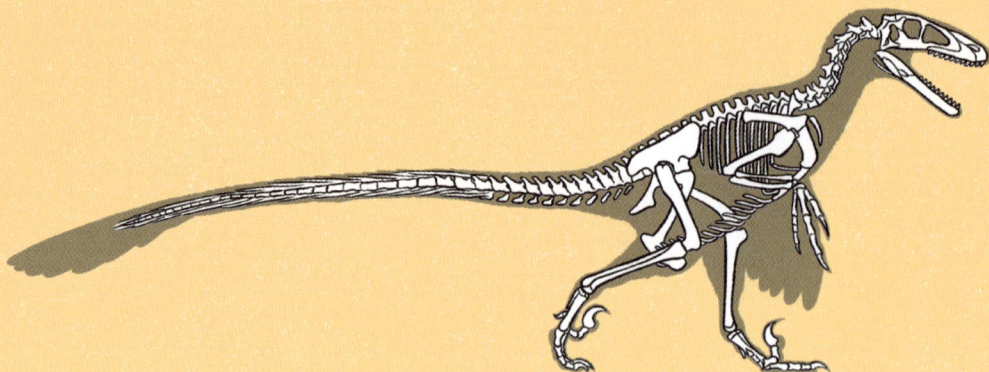

DEINONYCHUS

DYE·NON·EE·KUS

FIRST DISCOVERED: 1931 **RANGE:** USA

Not many dinosaur names evoke such instant imagery as that of *Deinonychus*, which means "terrible claw." Part of the group of theropods known colloquially as the "raptors" (and scientifically as the Dromaeosauridae), *Deinonychus* was a deadly hunter.

This 6-foot-tall (1.8 m) predator would use the 4.5-inch (11.5-cm) talon-like claw on its inner toe to bring down prey in the middle Early Cretaceous North America. It's thought that by using the claw to dig in and pin down their prey, *Deinonychus* could have followed up with sharp and effective bites to finish the kill. "Wings" of long feathers on the arms could have helped them stabilize themselves, too, especially if their prey was thrashing in an attempt to escape the dinosaur's clutches.

Considered an agile and lightweight hunter, the discovery of *Deinonychus* led to a shift in how dinosaurs were viewed by scientists and the public alike. The old view of lumbering monsters was replaced by a vibrant and fast-paced world of real animals, every bit as active as the world we see today.

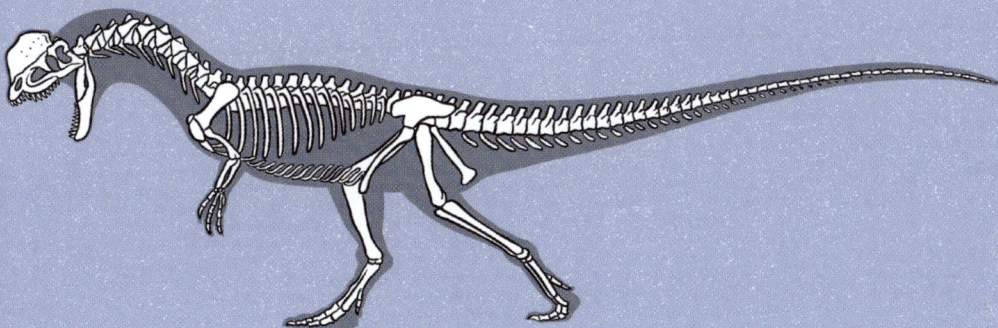

DILOPHOSAURUS

DIL-O-FO-SAW-RUS

FIRST DISCOVERED: 1940 **RANGE:** NORTH AMERICA

Possibly no other dinosaur is so consistently incorrect in its pop culture reconstructions than poor *Dilophosaurus*. Thanks to the *Jurassic Park* franchise, myths of venom-spitting and expandable frills have haunted this dinosaur for over thirty years.

Though an exciting idea for fiction, the real animal had neither of these features. However, that impressive double display crest on its skull was very much present in this fascinating predator that roamed what is now North America. These structures were probably brightly colored, functioning as elaborate display surfaces for competing dinosaurs.

The bizarre skull features don't stop there, as *Dilophosaurus* also possessed an interesting "notch" in its jaw. This shape may have made its bite less powerful but, combined with the long, sharp teeth, it was perfect for locking in struggling small prey.

At around 6.5-foot (2 m) tall and up to 23 feet (7 m) in length, *Dilophosaurus* was one of the first large theropods, and its size was unusual for the early Jurassic (though some similar-sized predators had appeared in the late Triassic). There's no denying dinosaur life could be full of strife, and one *Dilophosaurus* holds the unfortunate record for having the most bone injuries recorded in an individual dinosaur specimen.

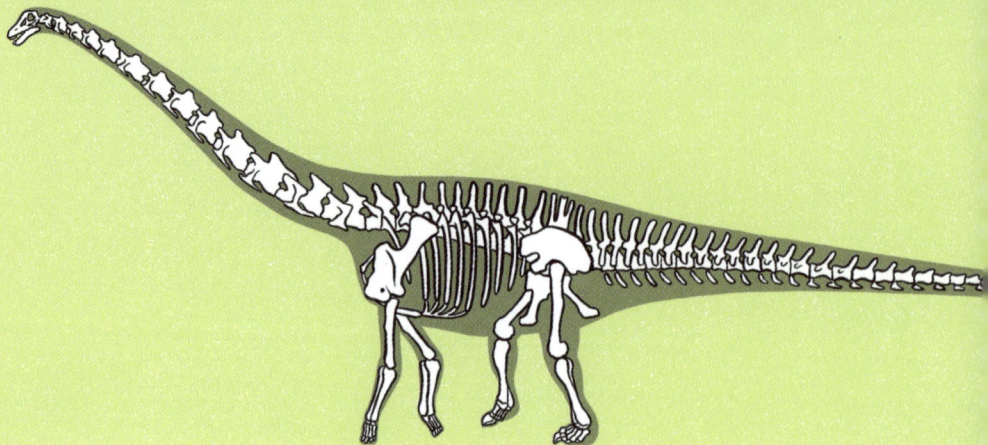

DIPLODOCUS

DI·PLOD·O·CUS

FIRST DISCOVERED: 1877 **RANGE:** USA

With a huge 26-foot-long (8 m) neck, four columnar legs supporting a massive body, and a tail stretching 33 feet (10 m), *Diplodocus* is the textbook image of a sauropod dinosaur. This giant lived across what is now North America 152 million years ago, and the enormous number of fossils that have been found prove it was a true dinosaur success story.

Given *Diplodocus* shared its world with many other sauropod dinosaurs, a large amount of research has gone into figuring out just how one environment could support so many giant herbivores. Studies on teeth shape and wear patterns have established the ideal food they evolved to eat, revealing that different sauropodomorph groups specialized in different plants. This diet splitting is known as niche portioning.

Diplodocus had thin and narrow teeth, often described as being peg-like. These were perfect for stripping leaves from branches, and were often lost and replaced (these dinosaurs having many more replacements than just the one set we humans have). Other sauropods had teeth better suited to tougher, woody vegetation.

A favorite addition to any museum hall, one particular *Diplodocus* specimen has become a dinosaurian celebrity. Based on a composite skeleton made up of several individuals, copies of "Dippy" have spread from the original in Carnegie Museum (Pittsburgh) to London, Paris, and Argentina.

Skin impressions found at Mother's Day Quarry in Montana have shown that *Diplodocus* had a variety of different scale types across different regions of its body. Ranging from 1/16–3/8 inches (1–10 mm) in size, these finds break the misconception that scaly dinosaurs had a single style of scale across their bodies. Other skin fossils from Wyoming have also revealed that some diplodocuses had small spines along their backs.

Various theories have circulated about the function of *Diplodocus*'s exceptionally long tail. If they were patterned at the end, it's possible that swishing it back and forth may have been used as a visual signal to other members of their herd. Some even thought it could have been used in audio signaling, generating a mini sonic boom when cracked like a whip, though this theory is likely an impossibility. Others think that the tail evolved to facilitate better herd management; instead of having to stop and look around to stay connected, these giants could keep together simply by staying in touch with the other herd members' tails.

ADAPTED FOR GRAZING

The *Diplodocus* skull was elongated and lightweight with large eye sockets and nostrils positioned on top. It had peg-like teeth only at the front, adapted for stripping leaves. The jaw joint was located far back, enabling a wide gape. Its skull also featured a distinctively high and narrow structure.

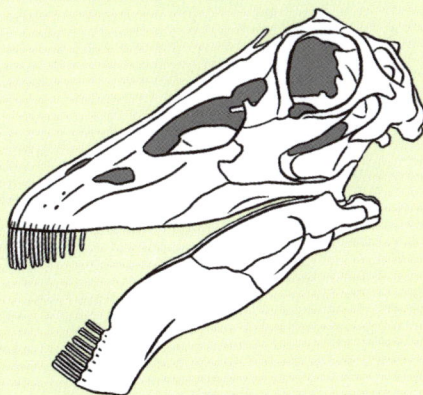

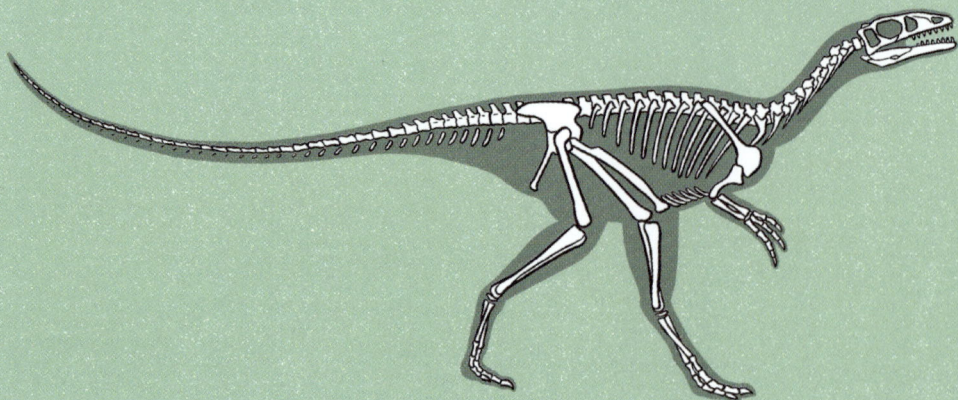

DRACORAPTOR

DRAY-COE-RAP-TOR

FIRST DISCOVERED: 2014 **RANGE:** WALES

This theropod discovered on the south coast of Wales was small by many dinosaur standards, around 10 feet (3 m) from head to tail. As such, it could be easily overlooked compared to its more famous relatives. Despite its size, however, *Dracoraptor* provides an invaluable record of a very important period of dinosaur history.

Dracoraptor lived about 201 million years ago, at the very start of the Jurassic period, when the dinosaurs were recovering from their first major challenge: the Triassic–Jurassic Extinction event. Fossils of predators of this time are exceptionally rare, so *Dracoraptor* received a significant amount of attention when remains of its skull were discovered in 2014, simply because they had been found at all.

The small pointed teeth of *Dracoraptor* were great for a generalist diet of small prey animals; insects, lizards, and early mammal relatives were all options, along with whatever carrion may have washed up on the shores of the tropical islands that made up Jurassic Wales.

Being from the land of dragons, *Dracoraptor* took the first part of its name from the Latin root for these mythical beasts.

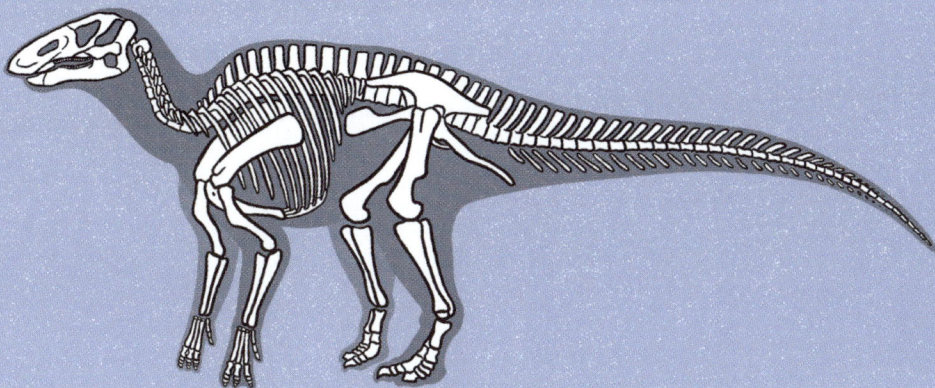

EDMONTOSAURUS

ED-MON-TOE-SAW-RUS

FIRST DISCOVERED: 1891 **RANGE:** CANADA AND USA

Another dinosaur named for an area of Canada, *Edmontosaurus* was a huge, 40-foot-long (12 m) herbivore from the Cretaceous period. The long skull of *Edmontosaurus* broadens at the end to a wide, flattened mouth, giving the group it belongs to the descriptive name of "duck-billed dinosaurs."

The mouth of *Edmontosaurus* was packed with densely concentrated units of teeth known as "batteries." Microscopic wear patterns on these teeth have led paleontologists to suggest that *Edmontosaurus* was actually a grazer, feeding on a revolutionary new plant that first appeared in the latter half of the dinosaur age: grass.

Regularly found within enormous assemblages of individuals (some sites are estimated to have contained fossils from over 20,000 dinosaurs),

Edmontosaurs are thought to have formed huge herds across the ancient landscape, reminiscent of the groups of buffalo that would walk the same lands 66 million years later.

One remarkable specimen of *Edmontosaurus*, nicknamed "Dakota," was preserved in extraordinary detail. Described as a "dinosaur mummy," Dakota showed scaly skin and nails, revealing that the fingers on the forelimbs of *Edmontosaurus* were covered in a singular "hoof-like" nail.

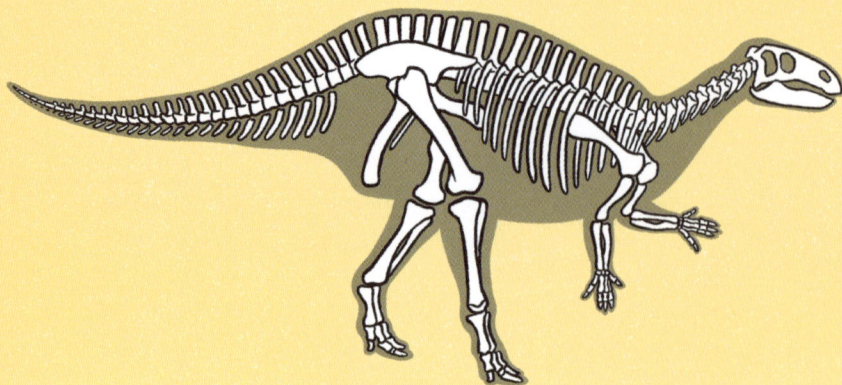

FUKUISAURUS

FOO-KOO-EE-SAW-RUS

FIRST DISCOVERED: 1989 **RANGE:** JAPAN

This striking 15-foot-long (4.5 m) herbivorous dinosaur from Early Cretaceous Japan had to wait fourteen years for its name as, despite first being discovered in 1989, no formal description of the fossils was published until 2003. In build, *Fukuisaurus* was similar in ways to a small *Iguanodon*; however, a key difference could be seen in the way it ate.

The fusing of the skull bones around the jaw of *Fukuisaurus* meant that the sideways chewing method seen in many other dinosaurs (known as pleurokinesis) would have been impossible for this animal. Instead, *Fukuisaurus* would have processed its food with robust vertical jaw motion.

The Fukui Prefecture of Japan has preserved a range of dinosaur species that lived alongside *Fukuisaurus*, including the sauropod *Fukuititan* and the feathered predator *Fukuiraptor*. In fact, about half of all the known dinosaur species discovered in Japan have come from Fukui. Each one has been found among the remains of fish and turtles, which would have lived in the rivers and lake beds that helped preserve the dinosaur skeletons so well.

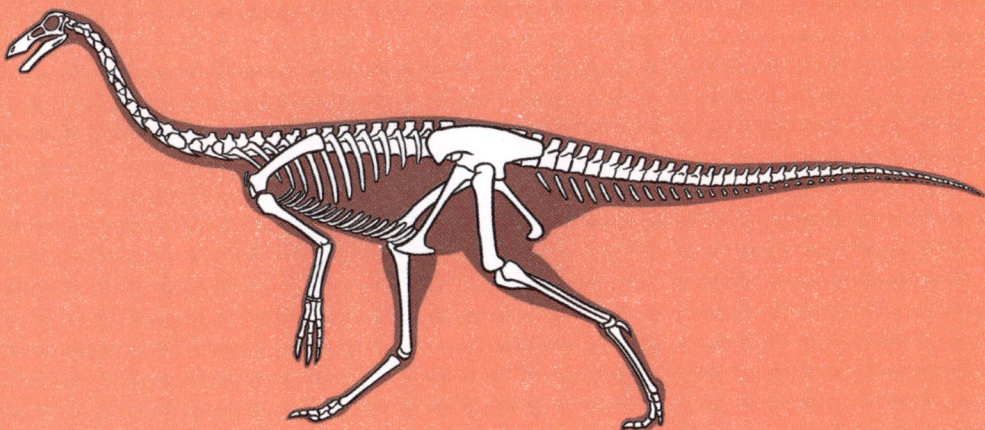

GALLIMIMUS

GAL-EE-MY-MUS

FIRST DISCOVERED: 1964 **RANGE:** MONGOLIA

Measuring 20 feet (6 m) in length, you could never call *Gallimimus* a small dinosaur, and yet it had a surprisingly light-looking build. Its long legs and reduced, rigid tail bring to mind the idea of a fast runner, and this is a reputation that has stuck with *Gallimimus* as a defining characteristic throughout its numerous film appearances.

Gallimimus is often depicted as living in groups, an idea supported by fossil evidence of individuals dying together during a harsh drought, as well as trackways of multiple *Gallimimus* moving in a unit.

First discovered during the celebrated 1960s Polish-Mongolian expeditions to the Gobi Desert, the habitat to which they belong, the ornithomimids ("bird mimics"), were previously known only from fragmentary finds, making three new skeletons invaluable in defining the species.

Beyond the powerful legs and flexible neck, original descriptions of the *Gallimimus* skull compared it to that of a goose, so bird-like was it in nature, completely lacking in teeth and covered in a keratin beak while alive. It is believed they were opportunistic omnivores, though predominantly herbivorous, swallowing small stones to aid in their digestion. Such a diet may even be the reason *Gallimimus* reached its impressive size.

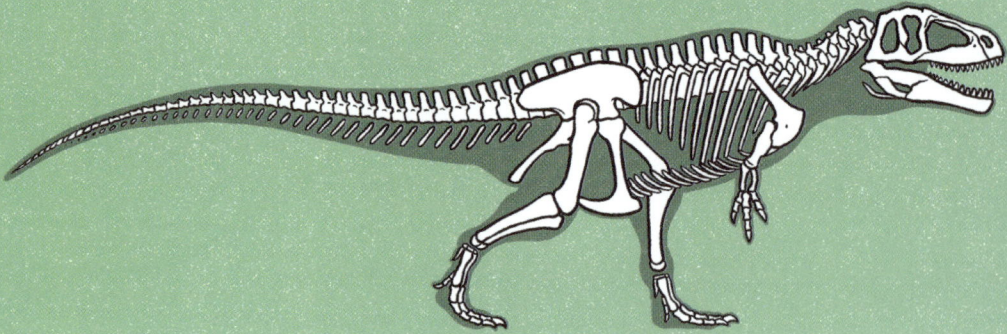

GIGANOTOSAURUS

GIG-A-NO-TOE-SAW-RUS

FIRST DISCOVERED: 1993 **RANGE:** ARGENTINA

During the Cretaceous period, what is now South America was a land of giants. The plants were harvested by the titanosaurs and, where prey is so large, the predators too can become enormous. As its name suggests, few could challenge *Giganotosaurus* in this respect.

Measuring 43 feet (13 m) in length, this giant carnivore was an apex predator, capable of using not only its mouth full of sharp teeth, but also powerful clawed arms to bring down dinosaur prey. With a distinctively bumpy textured head, *Giganotosaurus* resembled other giant theropods like the *Carcharodontosaurus*, though it lacked the crocodile-like spiked ornamentation it has often been portrayed with in pop culture.

Unlike the thickly built skulls of tyrannosaurs, which maximize their bite force, the skulls of *Giganotosaurus* and its relatives were slender and adapted for snapping shut rapidly. The skull of any predator must adapt to resist stresses, and *Giganotosaurus* may have evolved to handle this through a modification of its lower jaw, comprising a thick downward process of bone at the end. Though not exactly like ours, this feature is more or less the dinosaur equivalent of a having a particularly prominent chin.

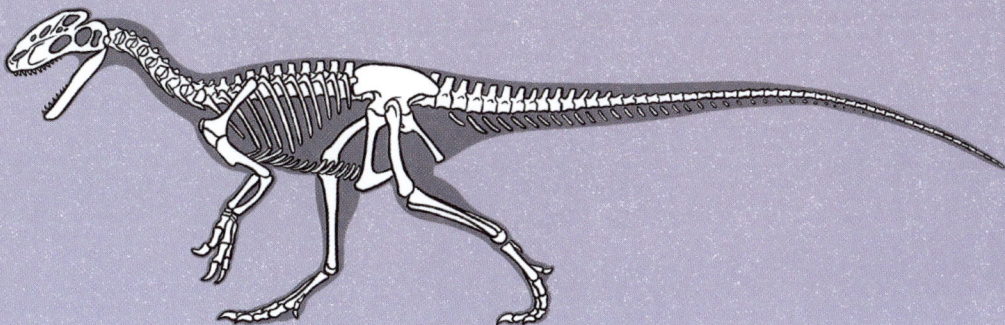

GUANLONG
GWAN-LONG
FIRST DISCOVERED: 2005 **RANGE:** NORTHERN CHINA

A little over 3 feet (1 m) tall, with long clawed arms and an elaborate display crest on its head, *Guanlong* looks very different to its more famous relatives. Surprisingly, this Jurassic dinosaur from what is now China is one of the earliest members of the Tyrannosauroidae family.

Unlike its short-armed ancestors, *Guanlong* was able to use its forelimbs for hunting prey. Modelling the performance of its claws has revealed they were ideal for piercing into flesh, and comparable to those of a puma today.

The bone structure of the impressive crest shows that it would have been very fragile, with large hollow spaces. Because of this, it is thought it would have been used only for display purposes, as it was likely too weak for any other function. This also means it would probably have been brightly colored, and may have been significantly different between males and females of the species.

Known only from the fossils of two individuals, these dinosaurs suffered an unfortunate fate of becoming mired in swampy mud, churned up by passing sauropod herds, trapping them like quicksand.

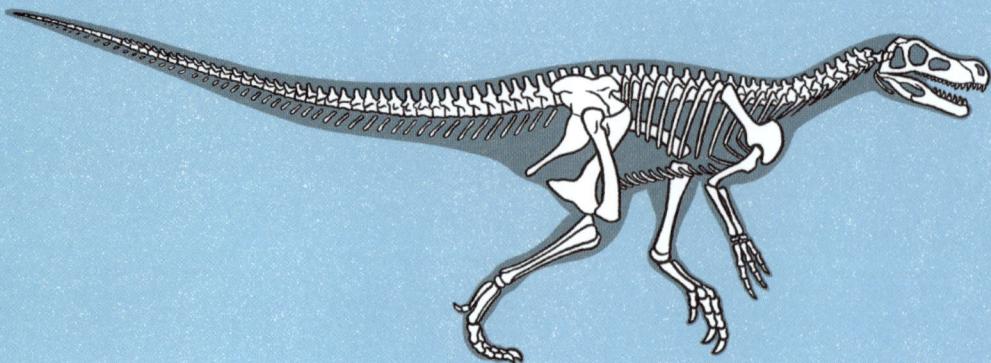

HERRERASAURUS

HE-RARE-O-SAW-RUS

FIRST DISCOVERED: 1959 **RANGE:** ARGENTINA

Dating to over 230 million years ago and one of the earliest members of the dinosaur family tree, *Herrerasaurus* shows many of the features that made it such a success story: legs held directly under the body instead of sprawling to the side like crocodiles; a lightweight, nimble frame; and a mouth full of teeth, perfect for an opportunist diet.

Such adaptability was key to surviving in the volatile climate of the Triassic period. At the time, these dinosaurs in what is now South America represented only a tiny proportion of the global ecosystem, but they would soon dominate the entire planet.

The rather boxy skull of *Herrerasaurus* is key to understanding its evolutionary legacy, as, although primitive, it shows a great degree of similarity to both the theropods and sauropods. This indicates that the split between the lizard and bird-hipped dinosaurs happened very early on in the group's evolution.

Most reconstructions of *Herrerasaurus* show it at a modest 13 feet (4 m) in length, but some individuals, known only from more fragmentary evidence, suggest that it could have grown much larger. The lower jaw bone of one such individual is twice the length of the best-known full skull.

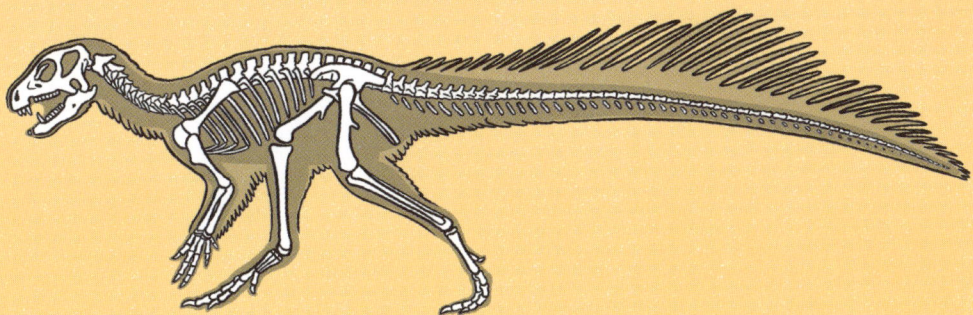

HETERODONTOSAURUS

HET-ER-ROW-DON-TOE-SAW-RUS

FIRST DISCOVERED: 1961 **RANGE:** SOUTHERN AFRICA AND LESOTHO

Vampiric-looking tusks set *Heterodontosaurus* apart. In fact, though dinosaur teeth in general are incredibly diverse and specialized for a wide range of unique functions across different species, unlike mammals, having such variations within the jaws of one individual is not common. Accordingly, the name *Heterodontosaurus* means "different-toothed lizard."

These teeth have led to speculation about what *Heterodontosaurus* likely ate. There is no doubt that the shearing tooth rows at the back of the jaw would help it cut up vegetation, the main portion of its diet. However, questions remain as to whether the lightly serrated tusks (and clawed hands) would have allowed catching small prey to add some protein to an omnivorous diet, or if they would be better suited for grazing and cutting through tough plant materials, like those of the muntjac deer.

From the early Jurassic, in what is now South Africa, *Heterodontosaurus* is a relative of those first to diverge to the ornithischian side of the dinosaur family tree, and has been an important study in understanding dinosaur breathing. The bizarre paddle-shaped bones associated with the sternum show a step in evolution to a more avian-like lung system.

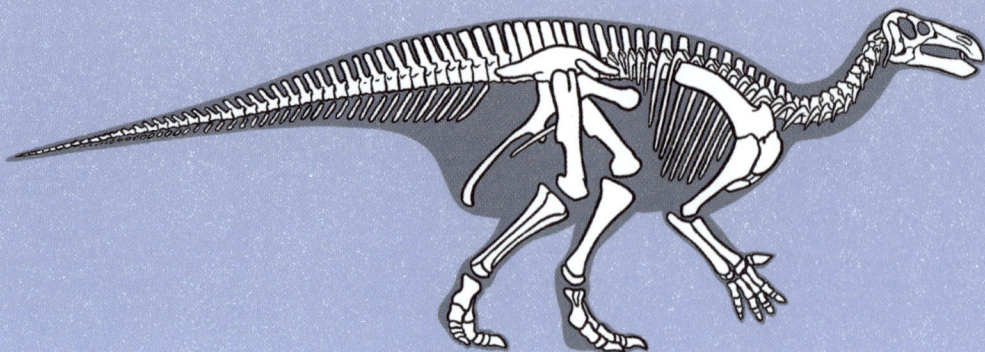

IGUANODON

IG-WAN-O-DON

FIRST DISCOVERED: 1820s **RANGE:** EUROPE

Influential in defining the dinosaurs to early paleontologists, *Iguanodon* was just the second dinosaur ever named. The story goes that British paleontologist Mary Mantell found two curiously large teeth in the rock spoils of a quarry in Oxfordshire, UK, in 1822. Her husband Gideon and anatomist Richard Owen thought they looked a lot like those of the iguana lizard, so gave them the descriptive name of "iguana tooth."

As scientists of the time tried to make sense of the giant bones that followed, *Iguanodon* found itself as the focus of some fairly wild reconstructions, from a lumbering reptilian rhino to a colossal tree-dwelling lizard.

Physical reminders of these efforts can still be seen in the giant sculptures of Crystal Palace Park, London.

A famous error was the initial assumption that a cone-shaped bone was a horn atop the nose, based on the living horned iguana. It wasn't until the "horn's" lack of symmetry and the shape of the base attachment were questioned a few years later that it was discovered to actually be a thumb-spike. This spike is often now depicted as being for defense, though it possibly served a function in digging up plant matter for the dinosaur to eat.

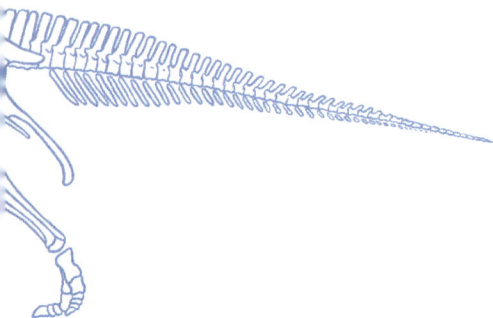

Those eponymous teeth were perfect for a diet of conifers and ferns, the fossilized pollen and spores of which have been found associated with the remains of *Iguanodon*. The front of the jaw was formed of a toothless beak that helped *Iguanodon* snip off bits of foliage before being ground up by the teeth at the rear.

Discoveries of multiple individuals together, including the presence of adults with several juveniles in Germany, has led to speculation that Iguanodons were social dinosaurs, travelling in herds to protect themselves and their young.

Iguanodon and its relatives were a successful group of dinosaurs through the Early Cretaceous, thought to have originated in what is now North America before spreading to Europe. The largest species of *Iguanodon* could reach huge sizes, measuring over 33 feet (10 m) in length. Finds of *Iguanodon* have been regularly reported in sites across Europe for two centuries now. However, in the many years since bursting onto the scene, as our understanding of dinosaurs has evolved, quite a few of the specimens first assigned to *Iguanodon* have been reclassified into a new genus, *Mantellisaurus*, as certain anatomical differences between the animals were identified.

BULK FEEDERS

Analysis of *Iguanadon*'s iguana-like teeth suggests that these dinosaurs were "bulk feeders." It is thought they spent less time actively searching for the best food in their environment, and instead ate whatever tough and fibrous vegetation they could find, thereby enabling them to maximize their nutrient intake by eating a lot.

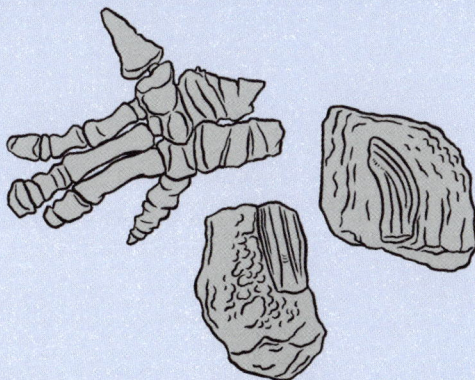

JOBARIA

JOE-BAR-EE-A

FIRST DISCOVERED: 1997 **RANGE:** NIGER

When huge bones were first found in the Sahara Desert, the nomadic Tuareg people of Niger believed them to be from a mythical giant, Jobar. The truth, that they came from a 60-foot (18-m) sauropod dinosaur that roamed Africa 160 million years ago, is even more incredible.

Moving such a long neck requires some special adaptations to the skeleton, and in the case of *Jobaria* and several other dinosaur species, this comes from a feature called bifurcated cervical ribs. These support structures are long processes of bone which are found underneath each neck vertebra, pointing backward and allowing for muscle attachment. Their size and splayed positioning mean that the neck muscles were diverging, giving the animal finer control over neck movements.

Dated to the Middle Jurassic, about 161 million years ago, the sauropods were still a relatively new group of dinosaurs when *Jobaria* lived. At 52 feet (16 m) in length, it was probably one of the largest terrestrial animals ever to have existed up until then, though its relatives would eclipse this by the end of the Mesozoic.

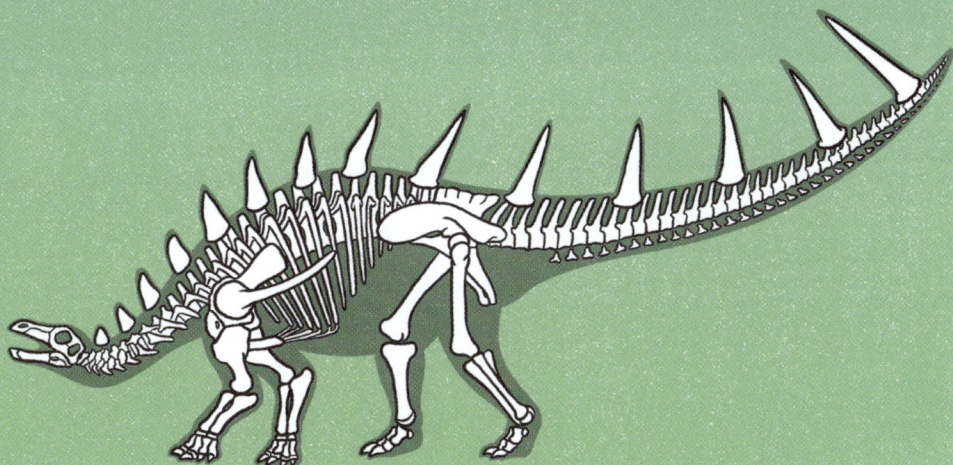

KENTROSAURUS

KEN-TROW-SAW-RUS

FIRST DISCOVERED: 1909 **RANGE:** TANZANIA

Stegosaur species are well known for the sharp spikes at the end of their tail, but *Kentrosaurus* took this a step further by having more spikes protruding from its shoulders. About two-thirds the size of its famous relative, its head was similarly positioned close to the ground for feeding on low-growing vegetation.

Unlike *Stegosaurus*, the plates along *Kentrosaurus*'s back transition to pointed spikes around the hip, rather than continuing down the full length of the dinosaur. Testing the strength of the tail spikes in action has shown that they would easily be able to damage the flesh of any attacking predatory dinosaur with continuous rapid movement.

Proving physical differences between the sexes in dinosaurs has always been difficult with nothing but their fossil bones to work from. However, some scientists think *Kentrosaurus* might have provided some evidence for it. Significant differences in size at the end of the femur have been speculated to represent such variation between sexes. Sadly, being isolated from other aspects of the skeleton means it's impossible to say for certain if any other diagnostic features may also have highlighted differences between males and females.

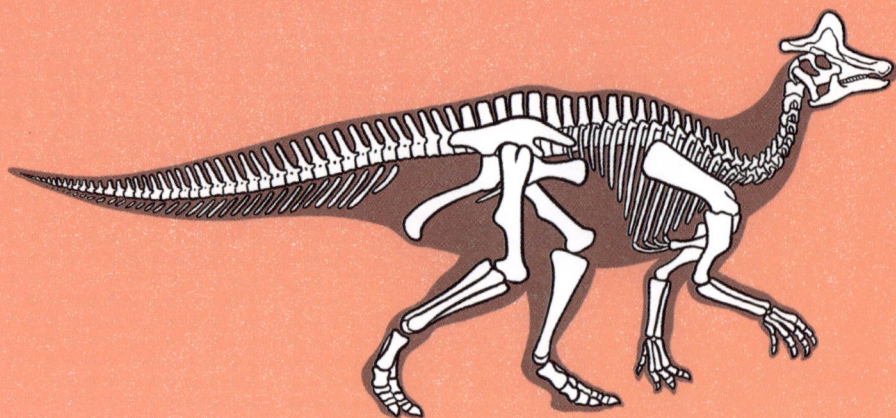

LAMBEOSAURUS

LAM-BEE-O-SAW-RUS

FIRST DISCOVERED: 1913 **RANGE:** CANADA

One of the largest known non-sauropod dinosaurs, *Lambeosaurus* is a hadrosaur (or duck-billed dinosaur) from Late Cretaceous Canada. Best known for its distinctive head crest, this ornate headgear was likely brightly colored, with a hollow structure that may have aided it in vocalizations.

Previously believed to have been other species of related dinosaurs, discoveries of juvenile lambeosaurs, which have much smaller crests compared to adult specimens, give an insight into this remarkable feature would develop as the dinosaur grew, only reaching its full size in adulthood. These crests were clearly important features for mate selection, as it is thought that their exaggerated enlargement evolved three times among the relatives of *Lambeosaurus*, completely independently of each other.

Unlike humans, who only have two sets of teeth throughout their whole lives, hadrosaurs like *Lambeosaurus* continuously replaced their teeth. Hundreds of small teeth united together to form what are known as "dental batteries." These units were used to grind up the tough plant material and, when worn, could be replaced easily and quickly. Each *Lambeosaurus* could have gone through thousands of teeth in its lifetime.

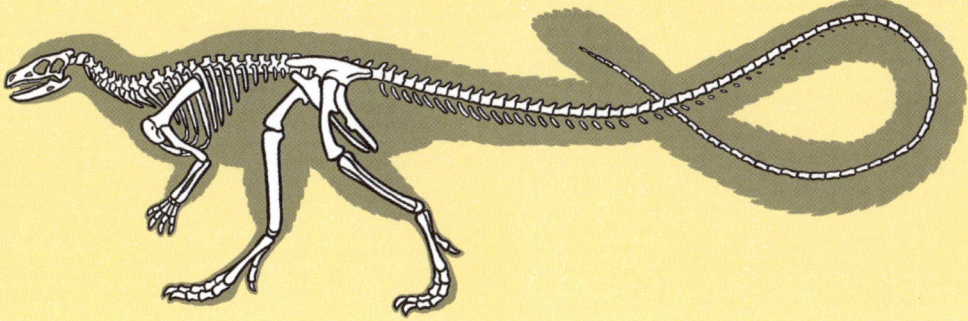

LEAELLYNASAURA

LEE-AL-EE-NA-SAW-RA

FIRST DISCOVERED: 1987 **RANGE:** AUSTRALIA

One of the most dramatic differences between the world of dinosaurs and today was the total lack of permanent ice caps at the poles. This meant that what now is an inhospitable desert of ice was once home to forests capable of supporting dinosaur life, such as *Leaellynasaura*.

When looking at a *Leaellynasaura* skeleton, the first thing you'd likely notice is the extreme span of the tail. Three times the length of the rest of the body, and flexible enough to move freely, the dinosaur might have used this almost like a fluffy scarf, wrapping it around its body to keep warm. After all, there may not have been permanent ice, but it was still significantly colder near the poles, especially for a small dinosaur (*Leaellynasaura* was only about 3 feet [1 m] in length).

Due to the large size of the eyes, it was proposed that *Leaellynasaura* had excellent sight, aiding it during the long, dark polar winters. However, this is more likely due to the fact that the first discovery was of a juvenile, which has relatively larger eyes compared to the rest of its skull.

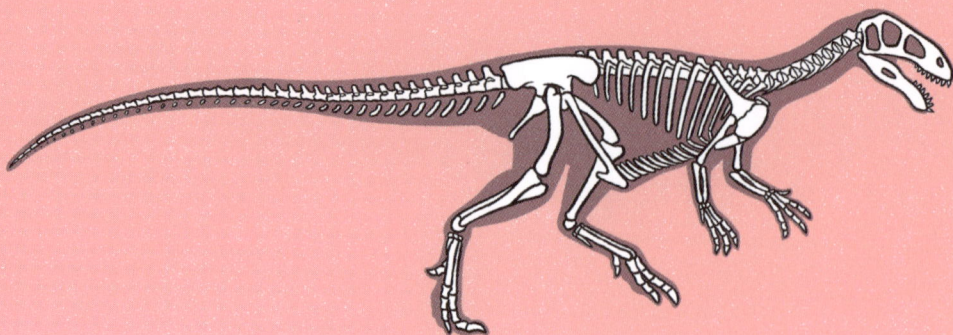

MASIAKASAURUS

MAS-EE-A-KA-SAW-RUS

FIRST DISCOVERED: 2001 **RANGE:** MADAGASCAR

At 6 feet (1.8 m) tall, this Madagascan dinosaur may be small in stature, but it makes up for its humble size with truly monstrous-looking jaws. The unique curved shape of the mouth means that the front teeth project outwards from the face, making them perfect for grasping onto small prey and locking them in place as they struggle.

The front projecting teeth are typical of theropod dinosaurs, with serrations to help cut through flesh, while the teeth to the back of the jaw are rounded at the apex. This implies multiple styles of feeding, perhaps even for getting through fruit. Either way, such localized differences in tooth shape is bizarre for a theropod dinosaur.

Thanks to an abundant fossil record, scientists were able to conduct studies on how fast *Masiakasaurus* could grow.

By examining the microscopic structure of the bones, it was determined that *Masiakasaurus* would have grown to full size in under ten years. This is a relatively slow growth compared to other theropod dinosaurs, possibly due to the stresses of living in a regularly arid environment. Staying small in size, and limiting the amount of food needed, likely also helped survival in these challenging conditions.

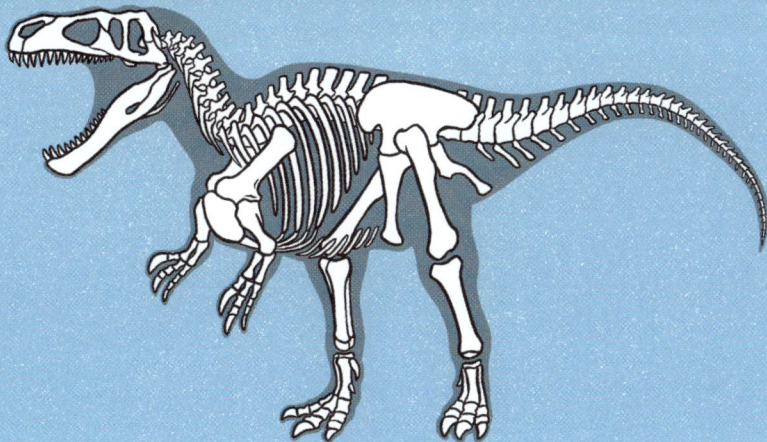

MEGALOSAURUS

MH-GA-LOW-SAW-RUS

FIRST DISCOVERED: 1676 **RANGE:** EUROPE

In February 1824, British theologian and paleontologist William Buckland announced a superb paleontological find from Oxfordshire, UK: the lower jaw and pointed teeth of some kind of "great lizard." He dubbed it *Megalosaurus*, and in doing so it became the first ever dinosaur formally named by science.

The first reconstructions depicted *Megalosaurus* as a cross between a lizard and an elephant, but in the two hundred years since then, our understanding of dinosaur anatomy has come on in leaps and bounds. We now know that *Megalosaurus* was in fact a bipedal carnivore, its forelimbs free and clawed to help bring down prey.

Despite having been discovered so long ago, there is still a lot we don't understand about *Megalosaurus*, with certain key aspects of anatomy as yet unknown to science. The exact snout shape, for example, as well as the total length of the animal remain a mystery, though most estimates put it at around 26 feet (8 m).

Once assigned to a litany of fossils across the globe, in reality, *Megalosaurus* had a much more limited geographical range, largely isolated by the inland seas of what is now Europe during the Jurassic period. Where *Megalosaurus* could be found, however, it stood as the apex predator.

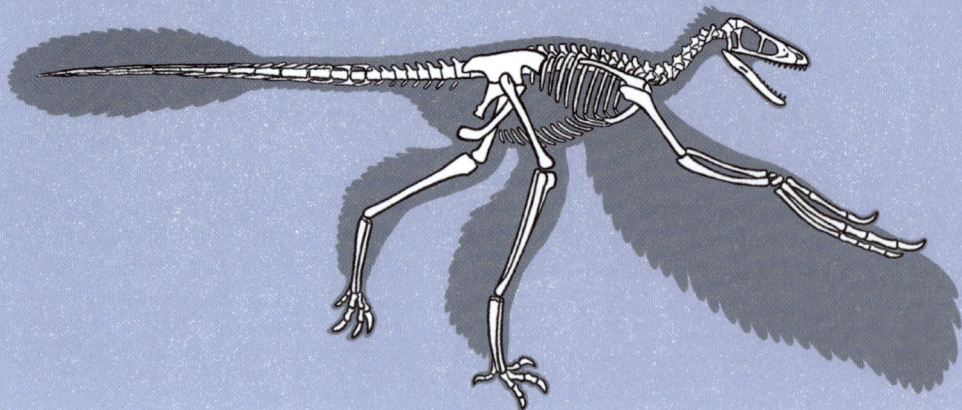

MICRORAPTOR

MY-CROW-RAP-TOR

FIRST DISCOVERED: 2000 **RANGE:** EASTERN CHINA

Having a small, lightweight body and long, asymmetrical feathers extending from both the arms and legs allowed *Microraptor* to do something remarkable: take to the air. *Microraptor* would not have been able to fly in the same way as true birds, and whether or not it was capable of powered flight, generating lift through flapping the wings, is still a matter of debate among scientists.

The microraptor's claws were well adapted for both predation and perching in trees, supporting the common idea that they would spend their lives in the canopies of Early Cretaceous forests, gliding between trees hunting small prey. Their eyes gave them acute vision, too, their binocular focus perfect for assessing distances and achieving precise landings.

Microscopic analysis of their fossilized feathers has shown that they were black and iridescent, reflecting a full rainbow spectrum as they caught the light. It is a feature we see in many birds today, used for elaborate displays of courtship and competition. Finding such an attribute in this Early Cretaceous theropod implies the dinosaur world may have been much more visually dazzling than in popular representation.

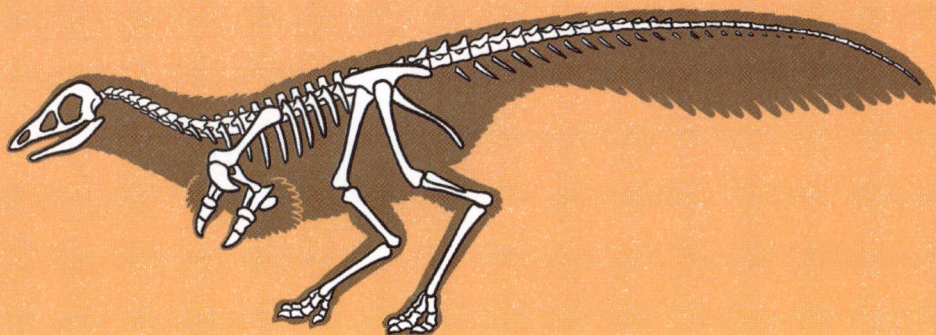

MONONYKUS

MO-NO-NIE-KUS

FIRST DISCOVERED: 1987 **RANGE:** MONGOLIA

T. rex is the subject of many jokes about its small, two-fingered arms, and *Mononykus* takes this even further, possessing tiny appendages with just one singular claw. Don't mistake size for weakness, however, as this arm could actually be described as dramatically over-engineered.

At a glance, *Mononykus* arms almost resemble those of a tamandua (tree anteater), and scientists believe this dinosaur could have been using them in much the same way. That 3-inch (7.5-cm) curved claw would have been perfect for tearing into insect nests. Examination of the musculature and range of motion of its arms shows they would have been able to forcibly flick back, meaning they could even have been good for burrowing.

Burrowing and the ability to reach hidden insects would certainly have been useful skills in the seasonal forests of Cretaceous Mongolia that *Mononykus* called home. A range of fossils at different developmental stages suggests that they lived here year-round too, no matter how challenging the conditions.

Though they lacked the special-ized mouths of modern anteaters, the jaws full of small, razor-sharp teeth give *Mononykus* away as a theropod, likely covered in feathers that could be fluffed up to regu-late temperature in harsh heat.

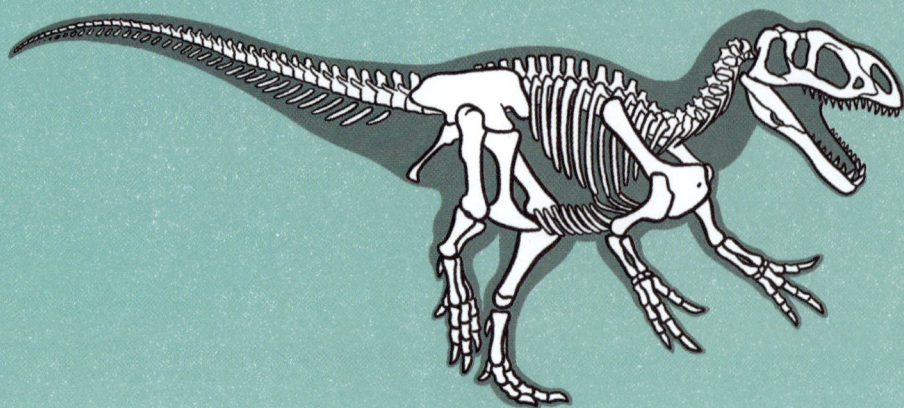

NEOVENATOR

NEE-O-VEN-A-TOR

FIRST DISCOVERED: 1978 **RANGE:** UK

Despite being 23 feet (7 m) in length, *Neovenator* was a lightly built hunter, likely living an active lifestyle. First found on the Isle of Wight, UK, in the 1970s, this theropod was integral in giving the island off the south coast of England its reputation as a European dinosaur hotspot.

During the Early Cretaceous, when *Neovenator* lived, what is now Europe was largely dominated by shallow seas and island ecosystems. This meant that, rather than the huge goliath predators of the present-day Americas, European hunters tended to be smaller and more able to adapt and take advantage of any opportunity that came their way. But with the only other large theropods in the area being the piscivorous spinosaurs (like *Baryonyx*), *Neovenator* was likely at the top of the dinosaur food chain at the time.

Close examination of *Neovenator*'s facial bones has revealed complex internal structures, meaning it was likely highly sensitive. Theories as to why this might have been vary, from improving the efficiency with which the carnivore could strip meat from a carcass to being for more intimate contact between individuals showing a surprisingly soft side to a group of animals often characterized as savage.

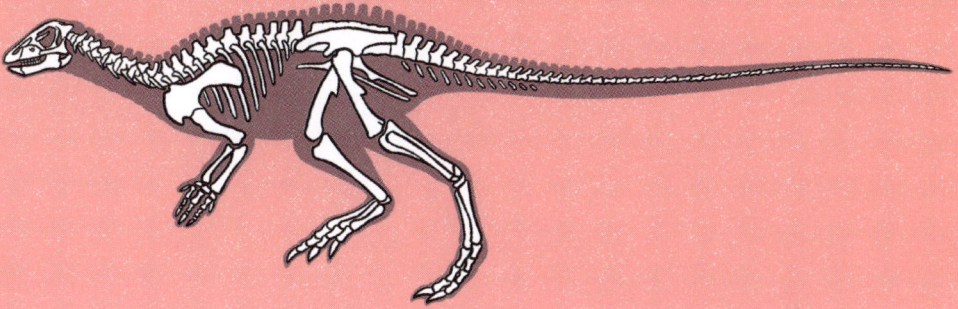

ORYCTODROMEUS

OR-EEK-TOE-DRO-ME-US

FIRST DISCOVERED: 2007 **RANGE:** USA

During the millions of years they ruled the Earth, the dinosaurs took advantage of every opportunity available to them. This includes several environmental niches that are not usually associated with the ancient reptiles. *Oryctodromeus* is an excellent example of this: a Cretaceous dinosaur from what is now North America that specialized in burrowing.

Analysis of the shape and size of *Oryctodromeus* shoulder and arm bones showed they were more robust than you would expect from other similar ornithopod dinosaurs. This is pretty good evidence that *Oryctodromeus* would be digging its own burrows, rather than inhabiting those excavated by other animals.

The burrows they dug were composed of winding tunnels leading to a main chamber where the animals would rest, comparable to those of large digging mammals today, such as hyenas.

The remains of *Oryctodromeus* adults and juveniles have been found jumbled together, indicating that the parents cared for their young after hatching, much like modern birds. While the burrows may have protected them from predation, the fact that they have been found like this at all implies these individuals were the victims of tragic burrow collapsing events, likely due to waterlogged soil.

51

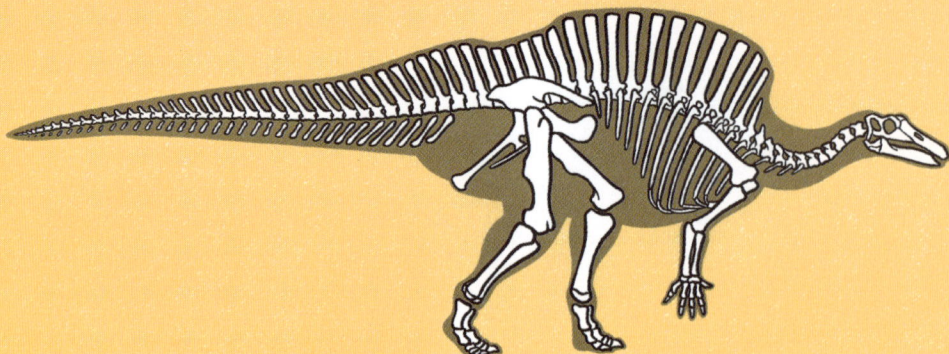

OURANOSAURUS

O-RAN-O-SAW-RUS

FIRST DISCOVERED: 1965 **RANGE:** NORTH AFRICA

Ouranosaurus was found in the midst of the Sahara Desert, so you might expect it to be a desert specialist, adapted to the harshest of conditions. Yet, 120 million years ago, in the early stages of the Cretaceous period, what is now a desert was then lush with vegetation and bisected by a shallow sea: the perfect environment to support groups of 2-ton (2,032-kg) herbivorous dinosaurs.

Starting near the shoulder and running down the full length of the body, the back of *Ouranosaurus* was dominated by a sail structure, extending over 23 inches (60 cm) at its peak. This marks it apart from other famous sail-backs, like *Spinosaurus*, whose sails stop before the tail.

By studying the structure of the bone and muscle attachment scarring, the best supported theory is that this sail played a role in regulating the dinosaur's body temperature. More extravagant ideas, such as it supporting a camel-like hump of fatty tissue, have also been proposed, though the lack of any soft-tissue preservation associated with *Ouranosaurus* mean this is more or less impossible to prove with the current fossil data.

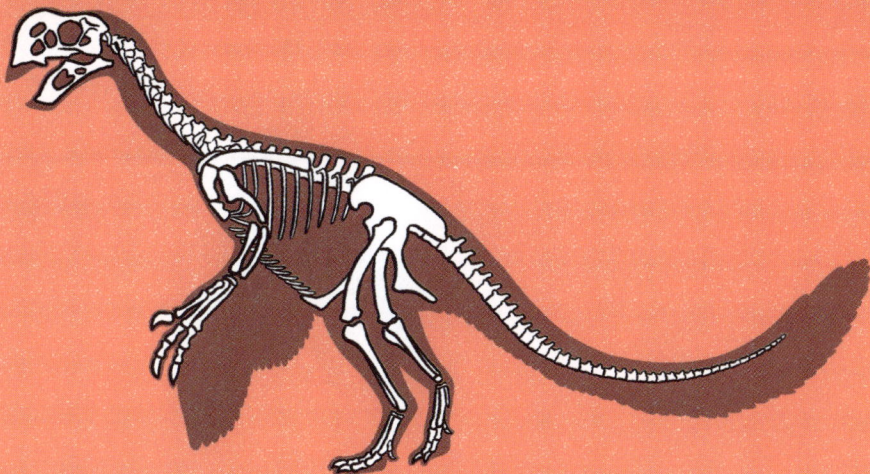

OVIRAPTOR

O-VEE-RAP-TOR

FIRST DISCOVERED: 1923 **RANGE:** MONGOLIA

The story of *Oviraptor* involves the most famous wrongful accusation in the dinosaur world. When this peculiar theropod was first discovered in Mongolia in 1923, it was supposedly found caught in the act of feasting on the eggs of another dinosaur. Thus it was given the name meaning "egg thief," and tarred with a reputation that would follow it for decades to come.

It wasn't until the 1990s that *Oviraptor* would be proven innocent, when it was shown that the eggs found alongside its remains were actually its own. *Oviraptor* wasn't hunting, but brooding, using its large feathered wings to protect its nest just as modern birds do.

Oviraptor's odd skull is often compared to that of a parrot, their toothless beaks being very effective at processing fruits, which likely made up a large portion of their herbivorous diet.

Despite being the first, *Oviraptor* is actually one of the least known from the group of dinosaurs it gave its name to, the oviraptorids, with only one confirmed full adult skeleton. *Oviraptor* itself is quite small, only about 6.5 feet (2 m) in length, but some of its relatives could reach four times this size.

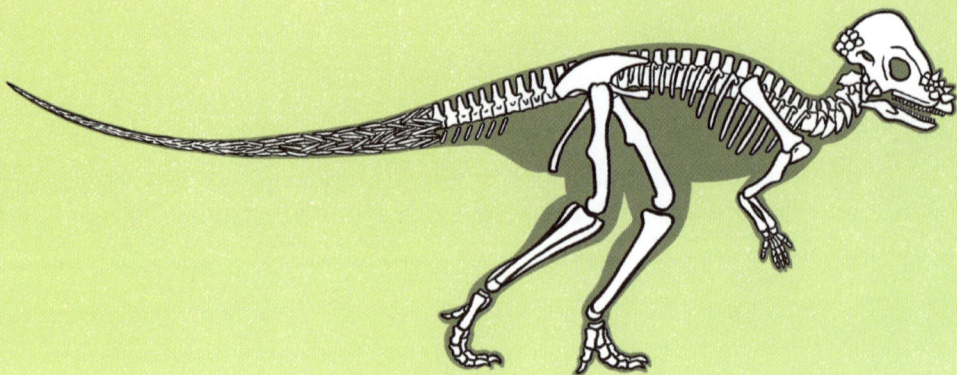

PACHYCEPHALOSAURUS

PAK-EE-SE-FAL-O-SAW-RUS

FIRST DISCOVERED: 1859 **RANGE:** USA AND CANADA

A dome of bone over 8 inches (20 cm) thick made *Pachycephalosaurus* an instant sensation when it was uncovered in 1931, and that feature has dominated the studies of this extraordinary dinosaur ever since. The immediate scientific reaction was that this was surely an extreme adaptation to impact absorption, for a lifestyle where butting heads was of the utmost importance.

The dinosaur as a whole was a little over 13 feet (4 m) in length and would have stood at around 5 feet (1.5 m) in height. A biped, the forelimbs of *Pachycephalosaurus* were highly reduced, but the legs were long and sturdy. They lived in what is now North America near the very end of the dinosaur era, in the Late Cretaceous.

Analysis of the typically broad teeth preserved in the original specimen gave a pretty clear consensus that *Pachycephalosaurus* was a herbivore, since they were perfect for ripping up foliage. However, a 2018 find that included the never-before-seen front sections of the jaw show bladed teeth typical of carnivores. It is therefore possible that *Pachycephalosaurus* was an omnivore, and would eat meat on occasion.

Despite the massive difference in appearance of their most famous members, *Pachycephalosaurs* are actually quite closely related to the ceratopsians, like *Triceratops*. The group to which they both belong, the Marginocephalia, are characterized by elaborate bone structures around the back edge of the skull.

Headbutting might be the obvious explanation for the huge dome of bone, but it isn't purely speculation. Evidence from traumatic lesions in the bone indicate that the scanned skulls were subjected to multiple traumatic impacts, likely from going head-to-head with a competitor of the same species.

Due to the massive forces involved, it has been suggested that *Pachycephalosaurus* would have used its head primarily on the flanks of its opponent, as seen in some modern antelope. The main argument for this is the rounded shape of the dome, meaning a direct collision would likely result in glancing blows and catastrophic accidents.

Whether or not *Pachycephalosaurus* would have used this battle-ram head as a defense against predators is harder to prove, but certainly an appealing idea. The same applies to the question of whether there was a difference in build between the males and females of the species. Unfortunately, answers to these questions are currently hindered by a lack of fossils.

MISSING FOSSILS

Pachycephalosaur fossils can be found across much of the northern hemisphere. Although many species are known, a complete skeleton hasn't yet been found. The thick skulls that made them famous are also the most resilient part of the skeleton, and so many individuals are known only from fragments of skull.

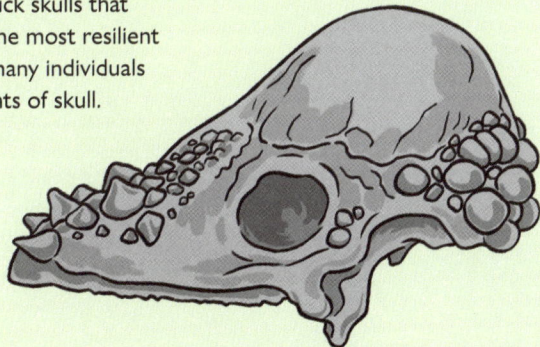

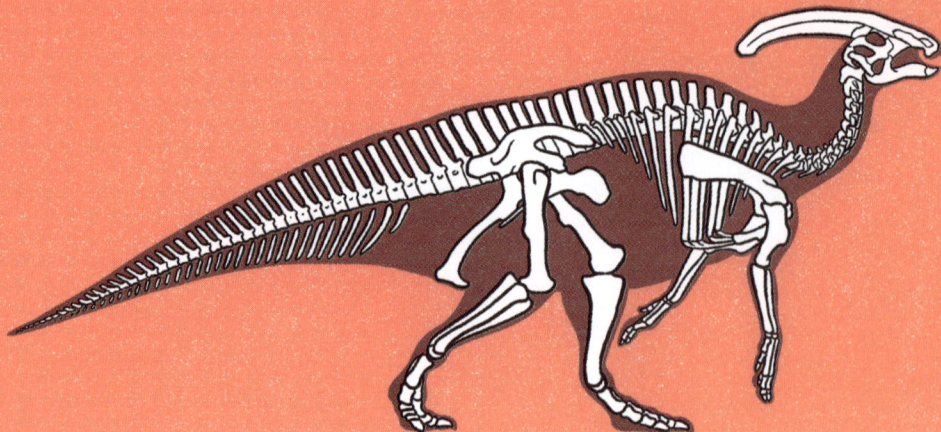

PARASAUROLOPHUS

PA-RA-SAW-ROL-O-FUSS

FIRST DISCOVERED: 1920 **RANGE:** USA AND CANADA

Possibly the best example of a dinosaur recognized by all but nameable by only a few is *Parasaurolophus*, thanks to its iconic headgear. That large crest, extending up to 5 feet (1.5 m) from the skull of this 30-foot-long (9 m) hadrosaur, was primarily a tool for communication, be that as a visual signal or for enhancing sound.

Internally, the crest is seen to have a series of air passages which would have acted as a resonating chamber, allowing the dinosaur to produce deep calls, perfect for communicating with other *Parasaurolophus* in herds. By modelling these structures, it has even been possible to attempt reconstructions of the sounds they may have made, though the accuracy of such reconstructions are difficult to verify without knowledge of how the soft tissue would have influenced the air flow.

The lack of any attachment structures or soft tissue associated with the crest implies that the crest was lacking any "sail" attachments, even though the dinosaur is sometimes depicted with these. It was, however, likely colorful or patterned as a display structure, and this idea is backed up by the well-formed eyes of *Parasaurolophus*, which suggest that vision was an important sense to it.

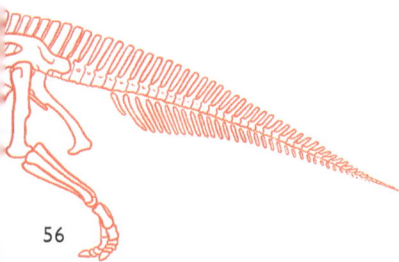

56

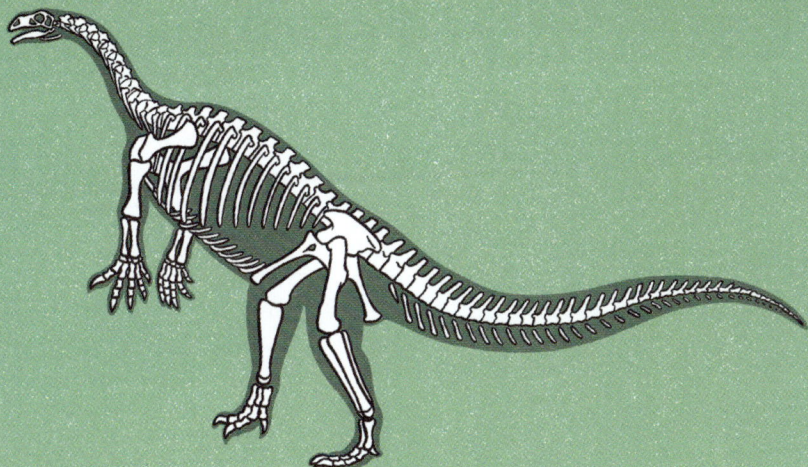

PLATEOSAURUS

PLAT-EE-O-SAW-RUS

FIRST DISCOVERED: 1834 **RANGE:** CENTRAL EUROPE

It didn't take too long into their reign for the dinosaurs to start getting very big, as the 33-foot-long (10 m) *Plateosaurus* plainly shows. A sauropodomorph from the Late Triassic of what is now Europe, it shows many of the features associated with the giants of later in the Mesozoic era; notably, that long neck for browsing usually out-of-reach foliage.

Despite being regularly depicted on all fours, *Plateosaurus* was not yet so big that it needed to do so, leaving its hands free to grasp branches as it fed. In fact, a study on how their hands were positioned showed that, because its wrists restricted hand movement, it would have been impossible for the dinosaur to have placed its hands flat on the ground.

Modelling *Plateosaurus*'s skull has established that it had a weaker jaw compared to that of the sauropods later in its family tree, but what it lacked in strength it made up for in closure speed. This evidence suggests that *Plateosaurus* may have been an omnivore and part of the evolutionary transition from the earliest, carnivorous dinosaurs to the sauropodomorphs, which would specialize as herbivores.

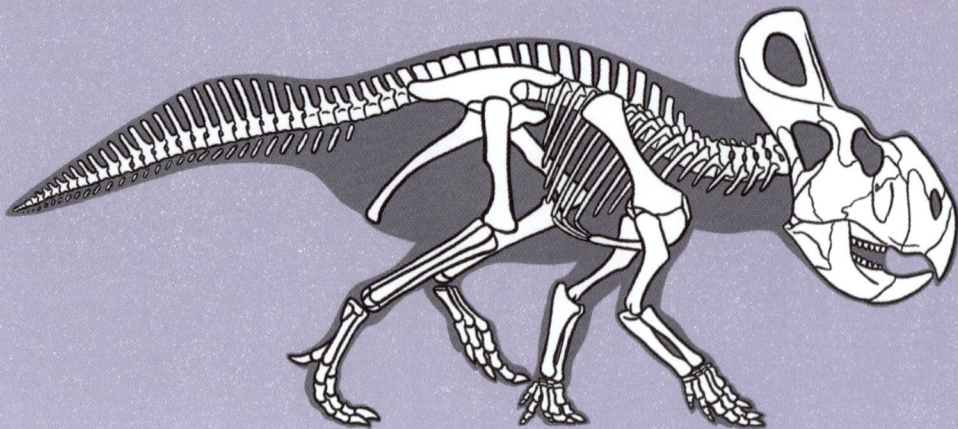

PROTOCERATOPS

PRO-TOE-SAIR-RA-TOPS

FIRST DISCOVERED: 1922 **RANGE:** MONGOLIA AND CHINA

Standing at just 20 inches (50 cm) in height, with a bony frill and no prominent brow horns, you might be forgiven for thinking *Protoceratops* is a kind of baby *Triceratops*. Though related and both part of the Ceratopsia, *Protoceratops* is from a separate branch of the family tree, and about 10 million years older. Seemingly living in herds, a huge number of *Protoceratops* skeletons have been uncovered, making them among the most studied of ornithischian dinosaurs.

Extensions protruding from the spine at the end of the tail appear to have formed a kind of paddle structure, which has been interpreted as evidence of *Protoceratops*'s somewhat amphibious nature. It's possible that the pointed beak and grinding teeth evolved for shearing off aquatic plants in the wetlands of Cretaceous Mongolia.

It seems that protoceratops employed a range of movement styles through their lives, as changes to the arch of the femur and width of the shoulder bones as they age suggest that juveniles were much more flexible in their motility than adults, even able to move about on just their hind legs rather than on all fours.

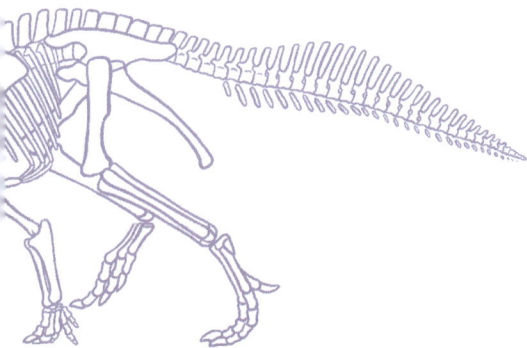

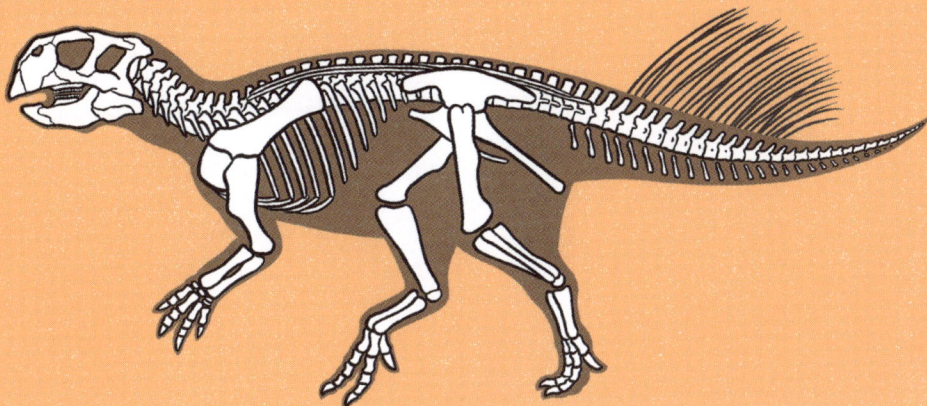

PSITTACOSAURUS

SITT-AC-O-SAW-RUS

FIRST DISCOVERED: 1922 **RANGE:** EASTERN ASIA

Named "parrot lizard" for its stout, beaked skull, this Early Cretaceous ceratopsian from what is now Asia is well known for an extraordinary fossil collection which has allowed scientists to reconstruct *Psittacosaurus* in astonishing detail. Indeed, hundreds of individuals have been uncovered and split into twelve different species, the most of any dinosaur genus.

Being among the first ceratopsians, *Psittacosaurus* lacks the large frills and horns of its more famous relatives (like *Triceratops*), but it did have an elaborate sail of feather filaments extending high above its tail, likely analogous to the tail display feathers seen in modern birds.

One extraordinary fossil of *Psittacosaurus*, known as the Frankfurt specimen, shows such detailed soft-tissue preservation across its whole body that some never-before-seen features have been identified, such as the fact that the shape of its scales varied across different regions of the body (cone-shaped near the shoulder, and quadrilateral on the tail).

Also identified was a crocodile-like cloaca, the single genital opening also found in birds. Seeing as this is present in their closest living relatives, it was long assumed to have been in dinosaurs too. And now, thanks to *Psittacosaurus*, there is hard proof to confirm it.

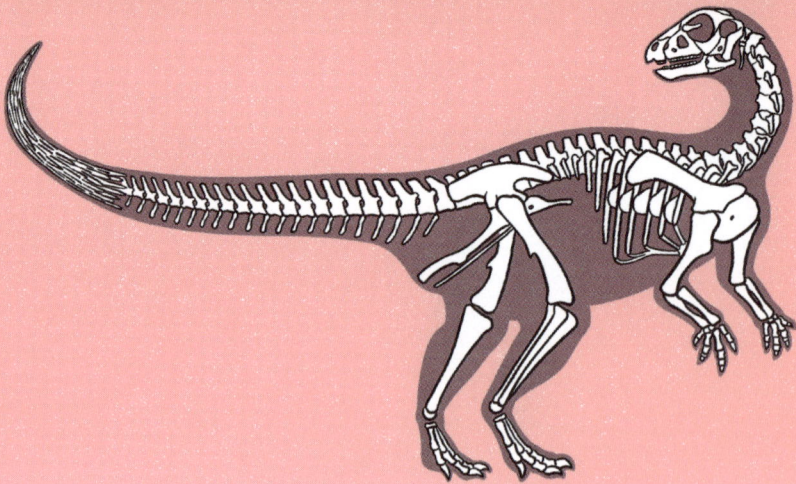

QANTASSAURUS

KWAN-TA-SAW-RUS

FIRST DISCOVERED: 1999 **RANGE:** AUSTRALIA

In science, it is not uncommon to name new discoveries after people, places, or even figures in pop culture. *Qantassaurus* is unique, however, in that it is the only organism named for an airline.

This small herbivore was another dweller of the Australian forests, living within the southern polar circle during the Early Cretaceous. At this time a great rift valley was widening in Australia, as Tasmania and Antarctica were splitting away, having been joined as one landmass throughout the Mesozoic to this point.

With only fragments of *Qantassaurus* jaw bone to go on, however, it is difficult to postulate what other adaptations this dinosaur may have had to survive the long harsh nights of winter.

Nevertheless, it can be seen that *Qantassaurus* had a stouter jaw than might be expected from similar species. This could be an adaptation for dealing with a wide variety of tough plant material, or it could be that they come from a juvenile dinosaur, not yet fully developed.

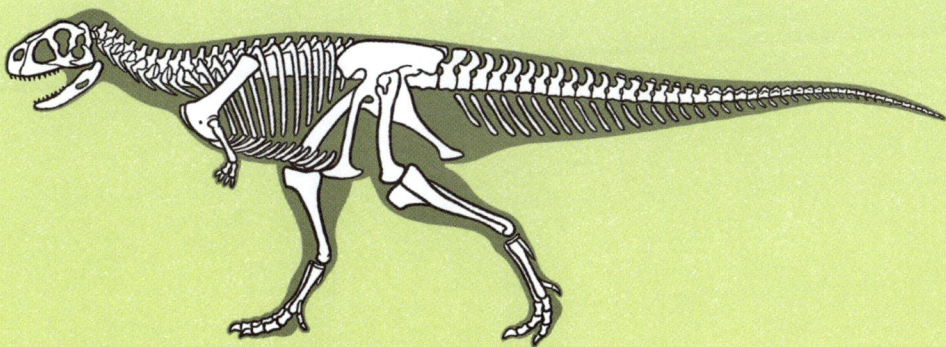

RUGOPS

ROO-GOPS

FIRST DISCOVERED: 2000 **RANGE:** NIGER

Known entirely from just a single skull from the Late Cretaceous discovered in Niger, paleontologists are lucky that *Rugops* has a particularly distinctive head, capable of revealing a lot about the whole animal.

A member of the abelisaurid family, these carnivorous dinosaurs have broad heads with a particularly flexible jaw joint, and bone with a textured external surface. While fossils of *Rugops* itself are limited, much more is known of their relatives, including *Carnotaurus*, allowing for an accurate reconstruction of the dinosaur's body, with small arms and long legs.

Whether or not an animal is fully grown can be estimated by whether the bones of the skull are completely fused. In the case of the one *Rugops*, it was indeed an adult, suggesting that, fully grown, *Rugops* would have stood at a similar height to the average human, and about 20 feet (6 m) in length.

The highly textured top of the skull suggests that *Rugops* would have had a covering of thick scales atop its head. What its purpose was is still uncertain, though one of the more popular ideas is that it might have been used in some kind of ritualistic headbutting.

SHUNOSAURUS

SHUN-O-SAW-RUS

FIRST DISCOVERED: 1977 **RANGE:** SOUTHERN CHINA

Sometimes in the dinosaur world, even being 33 feet (10 m) long and over 3 tons (3,048 kg) in weight isn't enough to ensure you're safe from predation. As such, some sauropods, like the relatively short-necked *Shunosaurus*, had extra protection.

The final few vertebrae at the tip of *Shunosaurus*'s tail are fused and inflated to form a bony club, tipped with two bone-cored spikes embedded in the skin. All of this could potentially have been a handy weapon in times of need, and may even have been used in competition with other members of the same species.

Questions have been raised, however, about the lack of reinforcement around the club. *Shunosaurus* is missing the "rigid handle" of fused vertebrae seen in the clubs of ankylosaurs, meaning that it may not have been able to whip the tail at full force without injuring itself. Enlarged neural canals around the tail have even led some paleontologists to propose that the club was in fact some kind of sensory organ, being used in a far gentler manner than first thought.

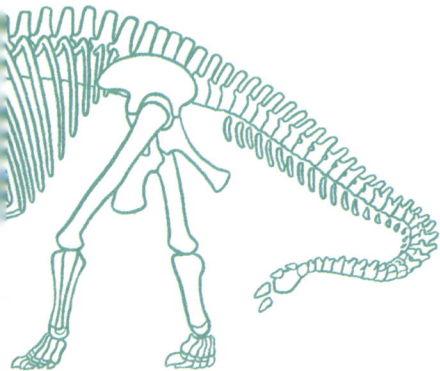

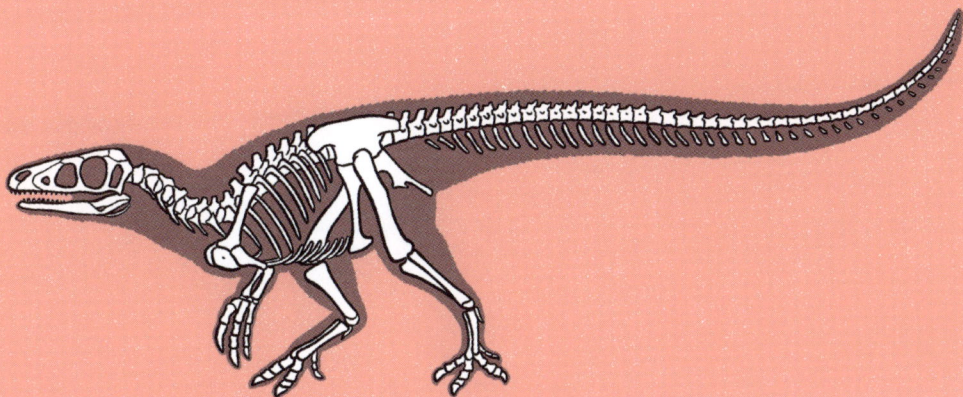

SINOSAUROPTERYX

SINE-O-SAW-ROP-TER-IX

FIRST DISCOVERED: 1996 **RANGE:** EASTERN CHINA

The link between dinosaurs and birds was speculated on for more than a century before a remarkable fossil from the Cretaceous, which was discovered in China, seemingly confirmed it. Unveiled in 1996, the dinosaur was named *Sinosauropteryx* and around the edge of the body were unmistakably the remains of feathers. This was the first feathered dinosaur ever found.

Not content with one monumental first, *Sinosauropteryx* claimed another in 2010, when it became the first dinosaur to have its color fully reconstructed. Thanks to microscopic structures in the feathers called melanosomes (explained on page 178), paleontologists were able to show that, in life, the plumage of *Sinosauropteryx* was ginger and white.

The dinosaur had a striped banding pattern on the tail, like a modern ring-tailed lemur. It may well have used it in the same way, held above the body to signal to others. The eyes were covered in a "bandit mask" of ginger feathers, similar to that of a raccoon, which might have aided in reducing glare from the sun while hunting small prey items in the forest clearings of ancient Liaoning.

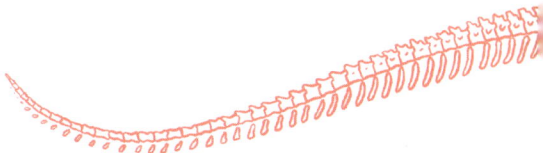

SPICOMELLUS

SPIC-O-MELL-US

FIRST DISCOVERED: 2019 **RANGE:** MOROCCO

Dating to the middle of the Jurassic period, *Spicomellus* is the oldest known of all the armored ankylosaur dinosaurs, though that is not the most intriguing thing about it. The defining feature of *Spicomellus* is a highly modified rib, with spikes fused to the upper surface.

Usually the armor is embedded within the skin and muscle, rather than directly on the bone. Even among the bizarre world of ankylosaurs, this is a unique bit of anatomy, not seen in any other vertebrate. However, despite the physiological differences, the purpose of the spikes was likely the same. In life, they would have been covered in keratin and acted in defense as a predator deterrent. Any theropod would be taking a huge risk in trying to take on such weaponry, even if, at about 10 feet (3 m) in length, *Spicomellus* was hardly a giant among dinosaurs.

Spicomellus lived in the forests of what is now Morocco, where it would have grazed on the low-lying vegetation. This same formation, the El Mers, has previously revealed one of the oldest known stegosaurs, making it an apparent hotspot for dinosaur evolution.

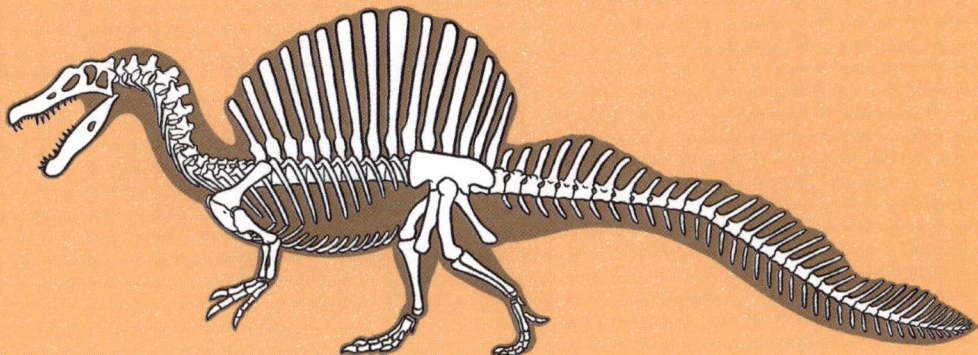

SPINOSAURUS

SPY-NO-SAW-RUS

FIRST DISCOVERED: 1912 **RANGE:** NORTHERN AFRICA

The largest meat-eating dinosaur that we know of is also one of the most controversial to reconstruct. What we do know is that *Spinosaurus* measured 46 feet (14 m) from snout to tail, and had a huge sail on its back; juts of bone extending from its vertebrae measured up to 5.2 feet (1.6 m) in height.

What precisely that sail may have been for is a matter of speculation, with theories ranging from regulating temperature, being a massive display banner, or helping it swim.

One thing that is clear about *Spinosaurus* is its diet. A crocodile-like snout; narrow, needle-like teeth; and huge hooking claws are all ideal adaptations to a life of fishing.

There's no doubt *Spinosaurus* lived around water, but the debate on reconstruction asks whether this creature was truly aquatic. Some argue it was fully evolved for swimming, with a wide paddle tail for propelling through the water. Others think it waded, like a titanic heron.

Due to limited evidence (many of the original fossils having been destroyed in World War II), it's possible that only more discoveries will solve the mysteries of this incredible dinosaur from Cretaceous-era Northern Africa.

STEGOSAURUS

STEG-O-SAW-RUS

FIRST DISCOVERED: 1877 **RANGE:** USA

It could be argued that *Stegosaurus*, *Triceratops*, and *T. rex* make up the "Big Three," the most famous dinosaurs ever discovered. With such a bizarre and fascinating body plan, unlike any animal alive today, it's not hard to see how *Stegosaurus* reached this level of celebrity.

Stegosaurus lived in what is now North America during the Jurassic period, about 150 million years ago, and has been making headlines since it was first discovered in 1877. Described by American paleontologist Othniel Marsh, *Stegosaurus* was named during the "Bone Wars" (Marsh's battle with fellow paleontologist E. D. Cope, and one of the most impactful and notorious scientific feuds of all time).

The function of the huge (some over 20 inches [50 cm] in height) plates along *Stegosaurus*'s back have fueled even more scientific arguments over the years, and remain somewhat of a mystery to this day. What is known for certain is that these plates were positioned in two staggered rows, rather than in pairs, and they were lined with blood vessels, providing oxygen for a coating layer of keratin tissue.

Among the many speculations, two theories seem to be the most widely

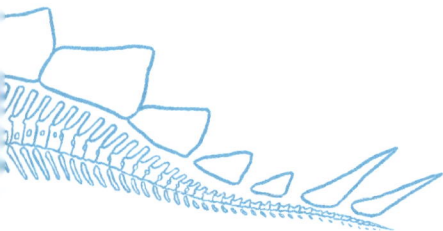

accepted by science. The first is that they may have aided in regulating *Stegosaurus*'s body temperature, pumping blood in to cool them down on hot days. The second is that they were used as display structures, brightly colored for recognizing individuals or attracting mates. The fact that their offset nature makes for a uniform visual banner on either side of the animal seems to lend support to this theory.

Stegosaurus's skull is small and relatively simple, only about 14 inches (36 cm) from nose to neck, compared to the total body length of the animal of 24.6 feet (7.5 m). Like many of the low-grazing herbivores of the dinosaur world, the front teeth had been entirely lost and replaced with a horned beak. Teeth are still present in the rear of the jaw, where they are small but perfectly evolved grinding surfaces for the tough Jurassic vegetation.

Finishing off the iconic silhouette of *Stegosaurus* were four 3-foot-long (1 m) spikes at the tip of the tail. These weapons could be used to deter predators, which there is at least one case of direct evidence for. A singular vertebra of the top predator of the time, *Allosaurus*, was discovered with an injury perfectly matching the size and shape expected from an impact with one of these spikes.

TAIL STRENGTH

Because nobody can be certain precisely how *Stegosaurus* might have swung its tail, it is hard to calculate the forces it could have generated. However, calculations based on a few educated assumptions have led to an estimated force of 450N, swinging at a speed of 20 mph (10m/s). While strong enough to damage the bone, as seen in the *Allosaurus*, hitting something so tough would likely have damaged the spike in the process.

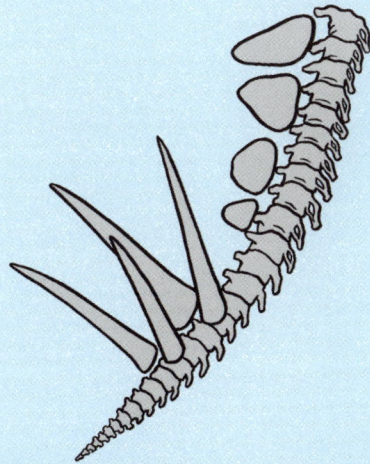

STYGIMOLOCH

STY-GIM-O-LOCK

FIRST DISCOVERED: 1990 **RANGE:** USA

Just the mention of the name *Stygimoloch* will prompt some dinosaur enthusiasts to object that it is no longer considered a valid dinosaur genus, and has instead been interpreted as a juvenile growth stage or different species of *Pachycephalosaurus*.

Certainly there are many similarities between them, physically and geographically (both being from the Late Cretaceous of what is now North America). The iconic dome skull is present in both, though it is slightly reduced in *Stygimoloch*, while the spiky protrusions around the edge of the skull are more elaborate. As such, it has been argued that *Stygimoloch* was still growing, that the dome would not be fully formed until the dinosaur reached sexual maturity and was ready to compete with it.

Debates such as these are common in paleontology. Deciphering family trees of living species is difficult enough, and relationships frequently change, even with access to genetic data. Doing so with only fragmentary fossils of long-extinct creatures is magnitudes more speculative. As more evidence is uncovered, opinions can and do change and, regardless of whether *Stygimoloch* is found to be a unique genus or not, the name will always be synonymous with the unavoidable uncertainty of paleontology.

STYRACOSAURUS

STY-RAK-O-SAW-RUS

FIRST DISCOVERED: 1913 **RANGE:** USA AND CANADA

Bearing a striking 20-inch-long (50 cm) horn above its nose, it is clear why *Styracosaurus* is regularly compared to a rhinoceros. But that was far from the only spike ornamenting this dinosaur. Two small horns protruded from its cheeks, and six more lined its impressive frill.

Best known from the famed Dinosaur Park Formation of Canada, *Styracosaurus* lived in the Late Cretaceous, a little under 10 million years before the extinction of the dinosaurs. As with other ceratopsians, *Styracosaurus* possessed extensive dental batteries of teeth evolved to process large quantities of vegetation for sustaining this 2-ton (2,032-kg) dinosaur.

Not only were the bony extensions from the frill impressive, they also seem to have varied highly between individuals. One example even shows asymmetry, with seven small extensions beneath the main horns on one side, and eight on the other.

When *Styracosaurus* has been uncovered, it is common to find the bones of multiple individuals, apparently traveling together as herds. It shared its environment with other ceratopsians, too, such as *Centrosaurus*, which also had one large nose horn. However, whereas in *Styracosaurus* this horn curves slightly back, in *Centrosaurus* adults the opposite is true, and it curves forward.

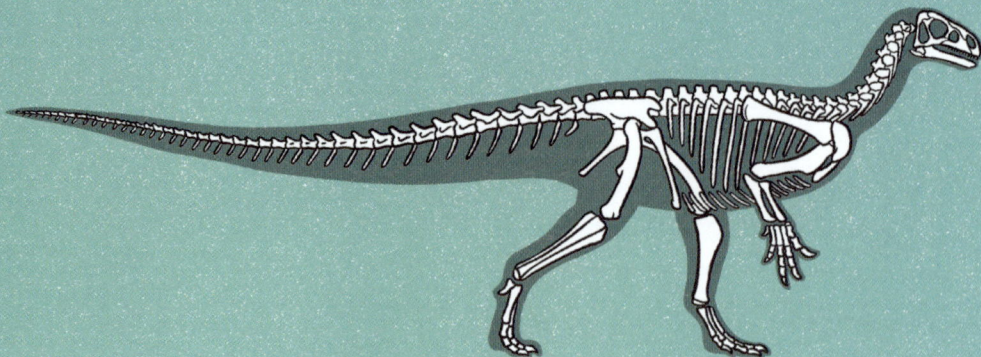

THECODONTOSAURUS

THEE-CO-DON-TOE-SAW-RUS

FIRST DISCOVERED: 1834 **RANGE:** EUROPE

In the early days of paleontology, mysterious bones coming out of a quarry in central Bristol, UK, were thought to have come from a small crocodile, and were excluded as a founding member of the Dinosauria. However, in the years since, *Thecodontosaurus*, the fourth dinosaur ever named by science, has proven integral to our understanding of early dinosaur evolution.

One of the first sauropodomorphs, muscular reconstructions of this sheep-sized dinosaur reveal that *Thecodontosaurus* was an agile biped, with front limb range of motion too limited to allow it to walk on all fours.

Examining the skull and braincase of *Thecodontosaurus* revealed that it had adaptations for maintaining head stability, which would be crucial for an animal that may have needed to make a speedy getaway from predators.

The leaf shape of the teeth, along with microscopic wear patterns preserved on their surface, indicate that *Thecodontosaurus* was a herbivore. This is a usual diet for a sauropodomorph, but particularly interesting in this case because such a diet was new to the dinosaurs during the Triassic, 208 million years ago, *Thecodontosaurus* being among the first species to make such a transition.

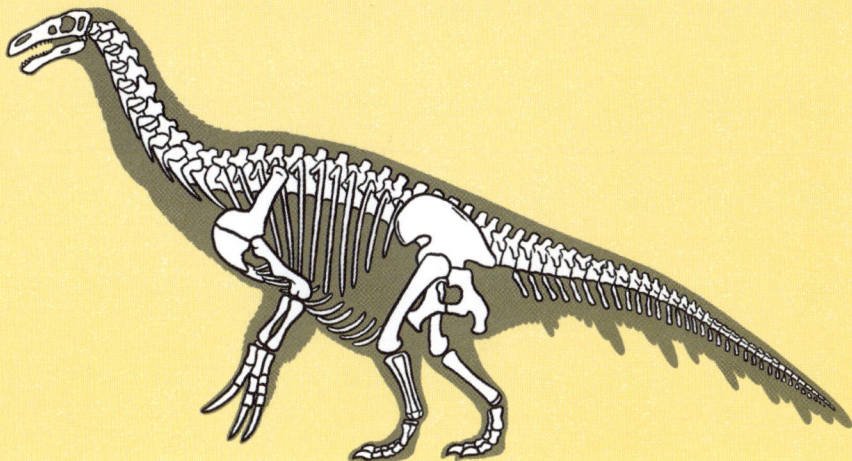

THERIZINOSAURUS

THE-RI-ZIN-O-SAW-RUS

FIRST DISCOVERED: 1948 **RANGE:** MONGOLIA

Each hand of *Therizinosaurus* bore three scythe-like claws up to 3.2 feet (1 m) in length. At first glance this may sound like the perfect weapon for a carnivore to take down prey, but this was not at all their function, because, as the rest of the body suggests, *Therizinosaurus* was a herbivore.

The head of *Therizinosaurus* was small and perched on a long and slender neck, while the legs were short and robust. Those legs and sturdy hips suggest a mode of life not usually associated with dinosaurs: sat down, upright, and grazing on vegetation like a modern gorilla. With evidence for a feathery body, a beak, and a hefty gut, perhaps the best way to think of *Therizinosaurus* truly is as the bird equivalent of a great ape.

Testing the mechanical strength of the claws suggested that they were most useful for hooking and pulling down vegetation, and not well suited to scratching and digging actions (the stresses generated on the top surface of the claws were too high). However, they could have had some function in defense, even if only due to their intimidating appearance.

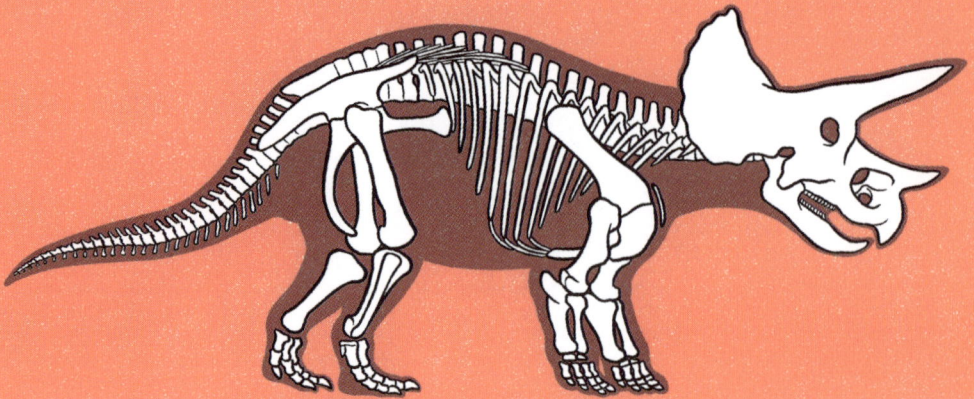

TRICERATOPS

TRY-SAIR-A-TOPS

FIRST DISCOVERED: 1887 **RANGE:** USA AND CANADA

One of the all-time iconic dinosaurs, *Triceratops* is unmistakable from the skull alone. A pair of 3-foot-long (1 m) horns mounted above the eyes, a smaller horn atop the nose, a beaked mouth, and at the back a huge frill of ornamented bone. It really is no wonder *Triceratops* still captures the imagination of the masses.

Ever since its discovery 130 years ago, it's been a common pop culture trope to see *Triceratops* facing off with *Tyrannosaurus*, using those horns to repel attacks. In fact, it is a situation that isn't too fanciful, seeing as these dinosaur species did live at the same time and place, namely at the very end of the Cretaceous period in what is today North America. We even have direct evidence of such encounters in the form of healed wounds on some *Triceratops* specimens, souvenirs of surviving an attack from a tyrannosaur.

Far more common injuries, though, are gouges, rakes, and punctures. These marks tend to be localized around the frills, and investigations show that they were almost certainly caused by the horns of other *Triceratops*, locking together in competition. It is a behavior seen repeated countless times in nature, the rutting of deer being perhaps the most well-known example.

Microscopic and chemical analysis of the bones from several different individuals reveal that they had healed from traumatic injury on multiple occasions, reinforcing the idea that such brutal competition was a recurring event in the lives of these dinosaurs.

Preservation of *Triceratops* endocasts (the space where the brain sits within the skull) have revealed more about how this dinosaur may have sensed the world around it. The olfactory bulb, responsible for smell, was relatively small, indicating that *Triceratops* didn't necessarily rely on smell to locate food.

The length of the cochlea (ear canal) is quite short, which means that *Triceratops* would have been adapted to hearing low-frequency sounds. This is pretty strong evidence that *Triceratops* would have used deep calls for communications.

By looking at the microscopic bone structure of *Triceratops* limbs throughout different stages of life, it has been calculated that they grew relatively quickly and continuously—a trait fairly common across the dinosaur family tree, many of whom were able to grow from eggs to multiton giants in just a handful of years.

HORN STRUCTURE

Today, we see the bone cores of *Triceratops*'s horns. However, in life they would have appeared longer and sharper, as the bone would have been covered in a sheath of keratin. If damaged, *Triceratops* wouldn't have been able to resharpen them like a cat can with its claws, so it likely took care not to damage its prized weaponry.

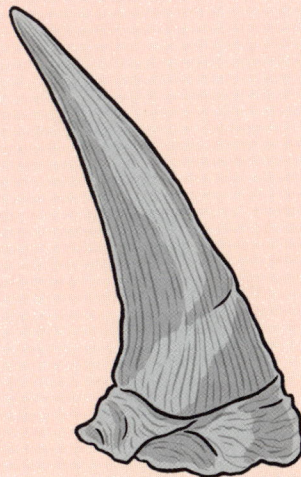

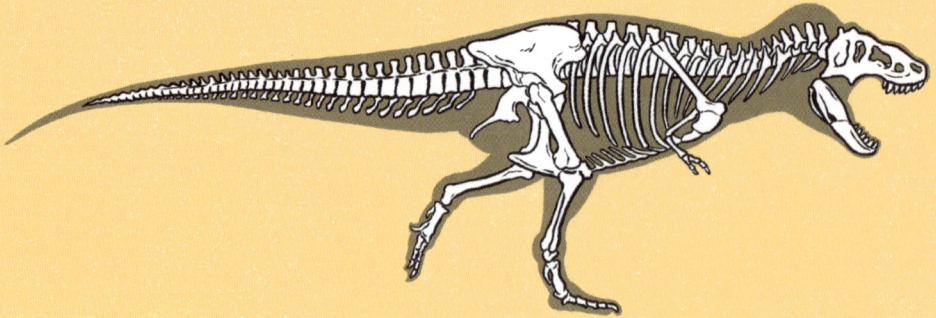

TYRANNOSAURUS

TY-RANN-O-SAW-RUS

FIRST DISCOVERED: 1874 **RANGE:** USA AND CANADA

A genus of giant carnivorous dinosaur from the Late Cretaceous (68–66 million years ago) found in North America. The most famous and popular of all dinosaurs, this apex predator stood 16.4 feet (5 m) high and 39.4 feet (12 m) long. The dinosaur's full name, *Tyrannosaurus rex*, is Latin for "Tyrant Lizard King," although it is most commonly known by the shorthand, *T. rex*.

The main weapon of *Tyrannosaurus* was its enormous head, which could deliver a bite force of 40,000N; the most powerful bite of any animal yet known. This would have allowed the animal to literally crush the bones of its prey. In order to balance out this heavy skull, the arms of *T. rex* were much smaller relative to its body size. To give an idea of just how much smaller, imagine an adult human with 5.1-inch (13-cm) arms. It is unknown if the arms of the *T. rex* served any practical function in life.

Fossils of *T. rex* have been found mostly in the USA, with many specimens discovered in northwestern states, such as Montana and South Dakota. Some of the more complete individual specimens have been named: the most

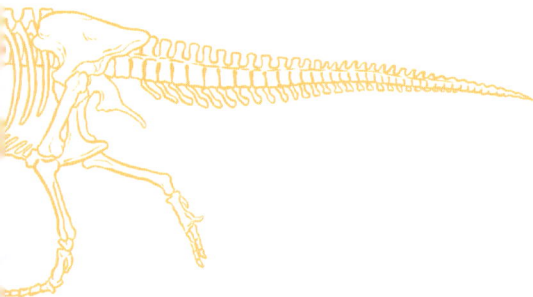

famous, dubbed "Sue," was found in South Dakota in 1990, and is displayed in the Field Museum of Natural History in Chicago. Despite the name, the sex of this remarkable specimen isn't known for certain—due to a lack of evidence, differentiating between male and female *T. rex* skeletons is a matter of educated guesswork.

SILVER SCREEN SENSATION

First discovered in 1900 by the American paleontologist Barnum Brown, *T. rex* was an immediate sensation and has since been featured in countless movie and TV shows as the go-to dinosaur star in mainstream media. Making its screen debut in the 1918 feature *The Ghost of Slumber Mountain*, *T. rex* is perhaps best known for appearing in the Jurassic Park series.

Although the status of *T. rex* as a hunter or a scavenger was previously debated, the evidence strongly suggests the former. Finely honed predatory features, such as front-facing eyes for better depth perception, as well as numerous fossils of its victims showing lasting scars from attacks, make quite clear that *T. rex* was indeed a hunter.

Among the last of the dinosaurs, *Tyrannosaurus* was the top of the dinosaur food chain when the asteroid struck the earth 66 million years ago.

A DINOSAUR CALLED SUE

In 1990, the American fossil hunter Sue Hendrickson made a historic discovery. Taking part in an expedition in the Cheyenne River Indian Reservation, South Dakota, she chanced upon the stunningly preserved remains of a *T. rex*. Now on display in Chicago, the eponymously named "Sue" is among the largest, most complete specimens in the world.

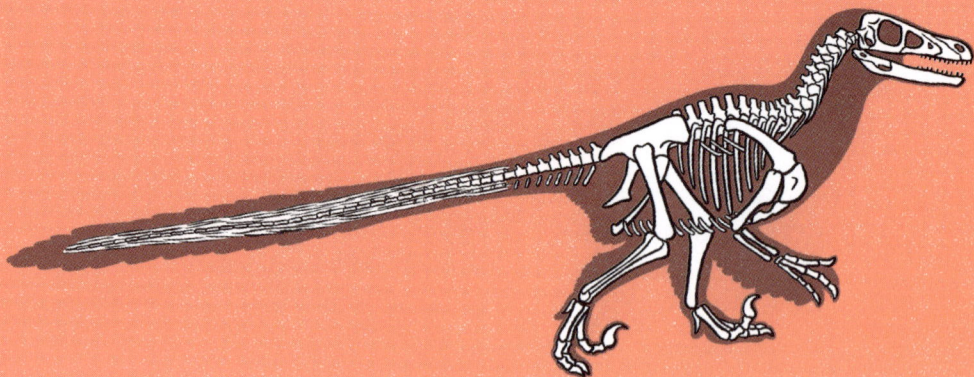

VELOCIRAPTOR

VEL-O-SI-RAP-TOR

FIRST DISCOVERED: 1923 **RANGE:** MONGOLIA

In science, the term "raptor" is used to describe the flying birds of prey, like eagles. However, thanks to a seeming omnipresence in pop culture, chances are that what most actually think of when hearing the word is a dinosaur. Specifically, *Velociraptor*. The real animal is quite different than what is usually seen on the screen, however.

Ever since *Jurassic Park*, *Velociraptor* has been thought of as a human-sized scaly predator, highly intelligent, with a killer claw on their foot. The first two statements have been definitively proven wrong by the fossils. In fact, *Velociraptor* was a small dinosaur, only about 24 inches (60 cm) in height, and covered in feathers. We know this not only because of the feathering of so many of their near relatives, but from direct evidence for *Velociraptor* itself in the form of feather quill attachments along the forearm.

Even the head shape of *Velociraptor* was different to how it is usually depicted. *Velociraptor* is distinctive among its relatives for having an elongated and narrow snout, with an upturn at the end. The teeth are sharp and leave no doubt as to its carnivorous diet.

That killer claw, however, was very real. One toe on each *Velociraptor* foot ended with a sickle-claw, reminiscent of the talons of a modern raptor. Being one of the first elements found, it was immediately glorified as a deadly weapon used for inflicting massive trauma on its prey, and, indeed, analysis of the claw has shown that its highly curved shape would have made it perfect for locking into prey and holding on. Surprisingly, it would have also been useful if the dinosaur wanted to climb trees.

One very famous fossil of *Velociraptor* even seems to show the claw in use. The incredible discovery, called the "Fighting Dinosaurs," was unearthed in Mongolia in 1971 and depicts a near-complete *Velociraptor* in combat with a *Protoceratops*, the two of them having died and been preserved perfectly during the fight. The arm of the *Velociraptor* is clamped in the mouth of the ceratopsian, which was apparently slamming its attacker to the ground when they were buried. The *Velociraptor* can also be seen using its claw to stab at the neck of the *Protoceratops*. It is doubtlessly one of the most incredible fossils ever discovered.

PACK HUNTER?

The intelligence of *Velociraptor*, and its ability to hunt in a pack, is still a debated matter. There isn't direct evidence for either trait, though they are both notoriously hard to test, and therefore rule out entirely. Studies of juvenile raptor teeth, however, show they may have significantly changed their diets as they grew, which is possible evidence against pack life. In pack animals, the juveniles eat prey killed by the adult members of the pack, meaning their teeth should be the same. That does not appear to be the case in raptors.

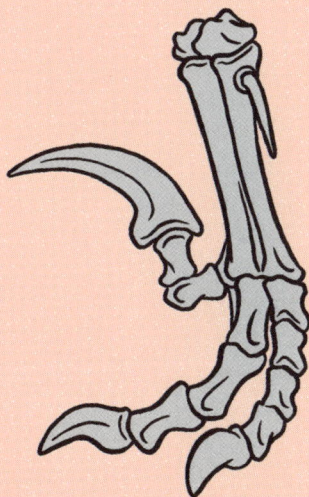

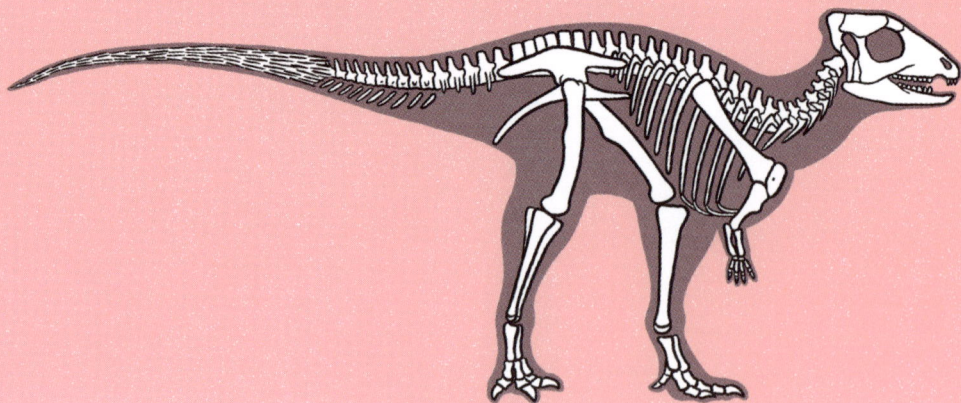

WANNANOSAURUS

WAN-AN-O-SAW-RUS

FIRST DISCOVERED: 1972 **RANGE:** EASTERN CHINA

Breaking the mold of the typical pachycephalosaur, *Wannanosaurus* is an exceptionally small dinosaur (maybe only 24 inches [60 cm] in length) and lacks the iconic thick dome skull of its relatives. Some have argued that it is merely a juvenile, but the mature features of the rest of the skull suggest that *Wannanosaurus* really was just a little guy.

Wannanosaurus lived near the very end of the dinosaur age, around 70 million years ago in what is now China. Known only from fragments, thankfully they are at least fairly distinctive. The jaw shows teeth which would be well adapted for a majority herbivorous diet, perhaps supplementing this with small amounts of meat should opportunity or need arise, while the back of the skull and extreme bending of the humerus cement its place on the pachycephalosaur family tree.

At the time of its discovery in the 1970s, *Wannanosaurus* was the second smallest dinosaur ever found. As such, even though the limited number of known fossils makes deep study of the animal difficult, it did help show that not all dinosaurs were Earth-shattering giants. The dinosaurs came in all sizes, great and small.

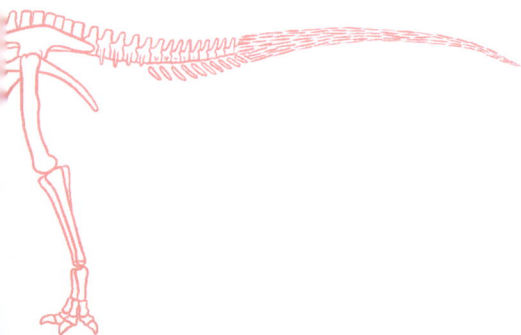

XENOCERATOPS

ZEE-NO-SAIR-A-TOPS

FIRST DISCOVERED: 1958 **RANGE:** CANADA

With a name that means "alien horned face," you'd be forgiven for thinking that *Xenoceratops* was bizarre-looking, even among dinosaurs. In truth, this name refers not to its appearance but its novelty in the area, being the only ceratopsian dinosaur found in the Foremost Formation, an important collection of fossil-bearing rocks that can be found across Canada.

At first known only from fragments of a skull, overlooked in a museum storeroom for over fifty years, several distinctive features of *Xenoceratops* could still be identified, including the classic long horns above the eyes. *Xenoceratops* lacked a nose spike, but the nose was not bare, and instead had an elongate flat structure running along it. Neither of these features are fully preserved in the fossil, but the fragmentary evidence strongly suggests their presence.

A few spiky nodules lined the edge of the frill, leading to the top where two huge protrusions capped off the crest. Despite the differences in shape compared to *Triceratops*, *Xenoceratops* was certainly using these features for the same purpose. The frill could be a display surface, while the horns and hardened nose structure would be locked together in conflict with rivals of the same species.

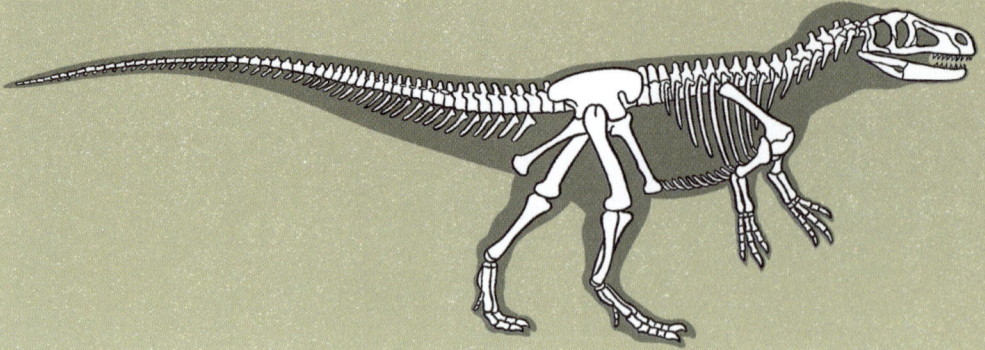

XUANHANOSAURUS

ZOO-WAN-HAN-O-SAW-RUS

FIRST DISCOVERED: 1979 **RANGE:** CENTRAL CHINA

As can often be the case in the study of dinosaurs, the memorable and bizarre early reconstructions that helped to make *Xuanhanosaurus* famous among paleontologists around the world would later be changed fundamentally by careful study of the evidence as science progressed.

When this theropod from the mid-Jurassic of what is now China was first uncovered, a great deal of attention was paid to the forelimbs. The humerus was very straight and the finger bones were larger than expected. This and more was used as evidence that *Xuanhanosaurus* walked on four legs, which would have made it the only known quadrupedal theropod.

Paleontologists have since agreed that this reconstruction is highly unlikely, namely because the distinctive wrist bones of theropods would have made it impossible for them to hold their hands flat to the ground and support any weight. Instead, the strong arms would have likely been used to slash at prey.

Another distinctive feature of *Xuanhanosaurus* is that it retains a functionless fourth finger on the hands. The first theropods had five fingers, reducing over millions of years down to three, which is the number seen in most later theropods.

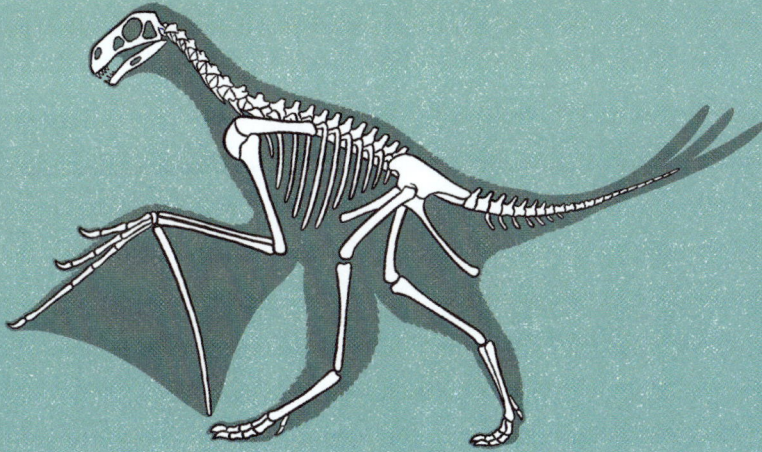

YI

YEE

FIRST DISCOVERED: 2007 **RANGE:** EASTERN CHINA

Surely a contender for the most bizarre-looking dinosaur of them all is *Yi*. This small (in size and name) dinosaur from the Jurassic of what is now China had arms that were long and functionally evolved for flight. These arms were not feathered, however, and instead flew with a membrane of skin, like the wings of a bat.

Supporting this membranous wing was a rod of bone, extending with a slight curve down from the wrist. No other dinosaur has such a structure, though it is curiously seen in some mammals (particularly in the ankles of bats). It isn't known for certain if these wings were capable of flapping flight, though seeing as *Yi* is missing powerful muscle attachment sites from the forelimbs, it is possible it was more of a glider.

Though the wing was composed of skin, the body of *Yi* was covered in feathers. Those around the head appear simple, but along the arm show a "paintbrush" style of feather which is much more unique. Being approximately less than 14 ounces (400 g) in weight and with a mouthful of sharp teeth, *Yi* most likely survived on insects and other small vertebrates.

YUTYRANNUS

YOU-TIE-RAN-US

FIRST DISCOVERED: 2012 **RANGE:** EASTERN CHINA

Having feathers is a trait we tend to associate with smaller dinosaurs, but *Yutyrannus* was certainly not small. In fact, at an estimated 26 feet (8 m) in length and weighing 2 tons (2,032 kg), it is the largest animal with confirmed feathers currently known to science.

The fossil feathers of *Yutyrannus* have been found from the neck, arms, and tail, suggesting full-body coverage. The chilly conditions of its native Liaoning Province of China in the Early Cretaceous were likely the reason for this plumage, with the 8-inch-long (20 cm) feathers giving *Yutyrannus* effective insulation.

The best skull found of *Yutyrannus* was flattened during fossilization, but comparisons to other tyrannosauroids enabled scientists to reconstruct its skull in three dimensions. Testing these models suggest a bite force of 2,500N, twice that of a jaguar (the strongest-biting big cat today). This still pales in comparison to the mind-boggling 40,000N of a *T. rex*.

That bite force was still more than enough for adults to tackle large prey, such as the sauropods they lived alongside, whose bones were recovered from the same quarry. Meanwhile, the thinner teeth of juveniles (who only had 10 percent of their adult bite force) would have been better for smaller local prey, such as *Psittacosaurus*.

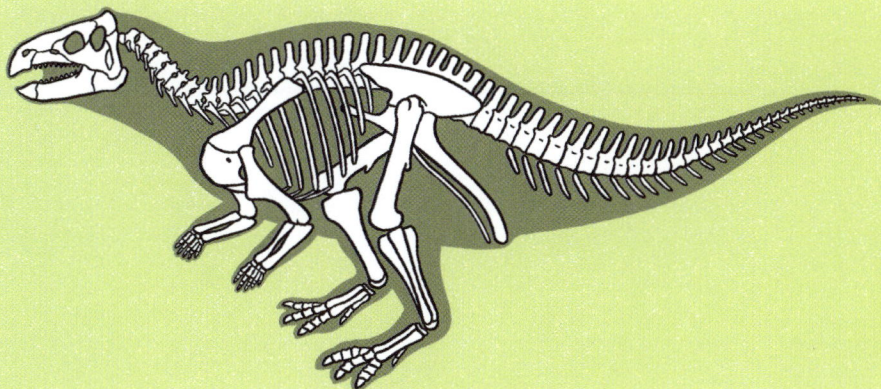

ZALMOXES

ZAL-MOX-EES

FIRST DISCOVERED: 1899 **RANGE:** ROMANIA

This Late Cretaceous dinosaur, discovered in what is now Romania, has long proved an enigma for classifications, its fossils having been reassigned on no less than four occasions between 1899 and 2003. Eventually it was given the name *Zalmoxes*, after a divinity that was described by ancient Greek historian Herodotus to have buried himself under the earth before re-emerging years later; a fitting name for a dinosaur.

The animal itself was a small bipedal herbivore, with a robust skull perfect for tackling tough plant materials. *Zalmoxes* was part of the famed Haţeg Island ecosystem, known for being the home of several dinosaur species that were far smaller than would be expected. This fossil assemblage is regularly looked at as a textbook example of "island dwarfism," whereby animals evolve smaller sizes due to the limited resources available to them.

Even now, however, *Zalmoxes* continues to cause grief to taxonomists, as, in 2023, a re-examination of the skull material of some specimens led to those fossils being ascribed to a different genus, *Telmatosaurus*. In this way, *Zalmoxes* is an excellent mascot for the unfortunate difficulties paleontologists often face trying to identify the remains of the long-extinct dinosaurs.

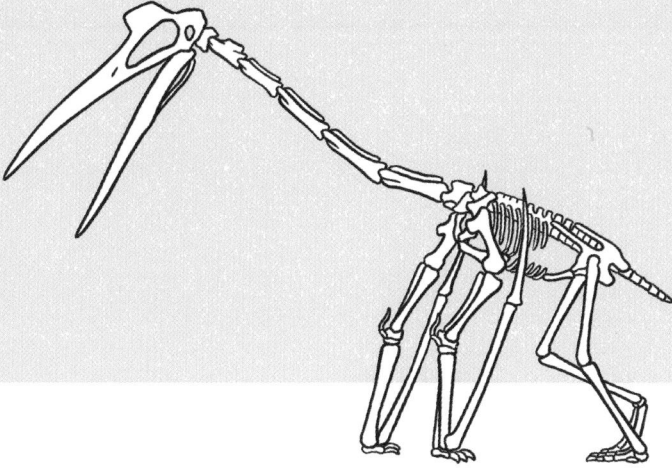

CHAPTER 2

PREHISTORIC BEASTS

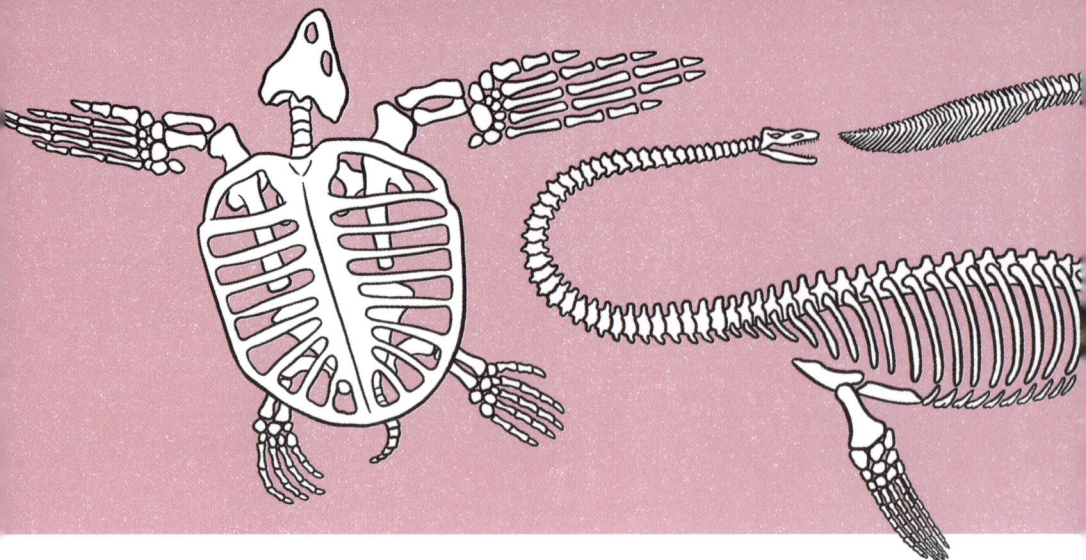

PREHISTORIC BEASTS

Dinosaurs loom so large in our imagination and understanding of the Earth's distant past that other creatures from the Mesozoic era are regularly overlooked by both the public and science. Despite often being lumped together under the heading of "dinosaur," however, the fossil evidence shows a world of boundless diversity at this time. Countless animal species evolved and were successful for millions of years, living in the seas, high in the skies, and at the feet of the dinosaurs.

IN THE SEA

If you were to dive into the Earth's oceans 230 million years ago, you would probably be familiar with many of the animals: the ray-finned fish, cephalopods (such as squid), and crustaceans (though true crabs would only evolve during the Jurassic). Sharks, too, could be found in the seas at this time, though it was not until near the end of the dinosaur age that any of the families we might recognize today would emerge. So, while the commonly repeated adage that sharks are older than dinosaurs and trees is technically true, those ancient sharks were very different than the ones we know today.

In the age of dinosaurs, it was not the sharks that sat at the top of the ocean's food chain, but the reptiles, who had also extended their dominance into the marine realm. The fish-like ichthyosaurs, paddled plesiosaurs, and sharp-toothed mosasaurs were the giant hunters of the Mesozoic seas, and will be covered more in this chapter.

IN THE SKY

Flying high above the dinosaurs during this time were the pterosaurs, whose wings were composed of a membrane of skin anchored by a greatly elongated fourth finger. These reptiles would evolve from small, long-tailed insect hunters to become the largest living creatures ever to fly.

It was during the Mesozoic, around 160 million years ago, that feathered birds took flight and began to compete for dominance in the air. As these birds evolved from within the dinosaur family tree, they too are technically dinosaurs. However, some diverged so much in behaviors and form that they have been featured here, rather than alongside the dinosaurs.

ON THE LAND

On land, then as now, the most diverse and abundant animals were the arthropods (the exoskeletal invertebrate group which includes all insects). Dinosaurs were the headline-grabbers of the era, but alongside them were representatives of the other major vertebrate groups too: the amphibians and the mammals.

It's a common misconception that mammals did not evolve until after the dinosaur age. These two groups did in fact live side by side, even though the former contained species far removed from any mammals familiar to us now.

In this chapter we will provide a snapshot of that extended diversity, touching on a few of the more iconic and notable groups and species. Some will no doubt be familiar, if only in passing from their depiction in prehistoric scenes, though others may be completely novel—hidden gems so often lost in the dinosaurs' shadow.

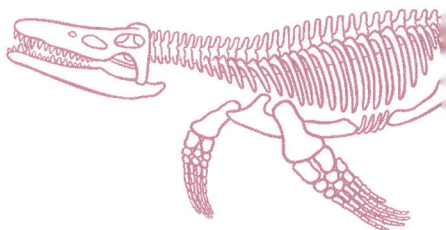

AMMONITE

A-MO-NITE

FIRST DISCOVERED: UNKNOWN (NAMED IN 1706)   **RANGE:** WORLDWIDE

The spiral shell of an ammonite is such an immediately recognizable shape that for many it is the first thing they imagine when they hear or read the word "fossil." Indeed, it is such an ingrained association that it can be easy to forget that ammonites weren't always mineralized remains, but were once a group of living creatures every bit as diverse and interesting as any invertebrate group alive today.

An ammonite's shell is divided into chambers, and in life each would be filled with fluids and gases which helped regulate its position floating in the water column. The main body of the creature itself inhabited just the final chamber, and would in many ways have appeared like a modern squid. Tentacles would have been used for grasping prey and transferring it to the hardened beak mouth (called an Aristotle's Lantern). Large eyes would have helped them locate their prey.

Their shells were composed of the tough mineral aragonite, and, like many modern sea shells, had a mother of pearl layer along the inside. Further protection came from plates called aptychi, set near the opening of the shell, which may have been able to cover up the soft body if retracted into the shell when the ammonite was faced with a predator.

Ammonites thrived right through to the very last day of the Cretaceous, becoming victims of the same cataclysmic event that spelled the end of the dinosaurs. It was once believed they were on the decline well before this extinction event, but new research suggests this wasn't the case, and if not for the asteroid strike, they may well still have been swimming in our seas to the present day.

A hugely variable group, some species were adorned with spines for further protection, and ridges called keels were common, similar to those of ships for helping them move through water. Sizes could range from only a few millimeters across to over 6 feet (2 m)—*Parapuzosia* of the Cretaceous, for example.

Most of us are familiar with the classic ammonite spiral, but some ammonites were much more bizarre. The heteromorph ammonites displayed two distinct shell regions, with a spiral top and a much straighter body chamber extending far below. One incredible genus, *Diplomoceras*, had a shell which, remarkably, resembled a paperclip.

SWIMMING SPEEDS

Much like modern squid, ammonites moved themselves through the water with a system of jet propulsion. Nobody is certain how fast ammonites could swim, but bumps and ornaments on the bulky shell would have greatly increased drag, so even classic ammonites were likely fairly lethargic movers. However, thin and bladed species, like *Oxynoticeras*, would have swum more quickly.

ARCHAEOPTERYX

AR-KEY-OP-TER-IX

FIRST DISCOVERED: 1861 **RANGE:** GERMANY

Among the earliest known birds, *Archaeopteryx* lived in the mid-Jurassic (between 163.5 million to 145 million years ago). Initially described from a single fossil feather discovered in a stone quarry in Solnhofen, Germany, in 1861, a full skeleton was uncovered later that same year.

This skeleton showed clear features of both birds and reptiles, suggesting an evolutionary link between the groups just two years after Darwin had published *On the Origin of Species* outlining the theory of natural selection in 1859.

Much of *Archaeopteryx* is bird-like, but the skull is very reptilian, complete with bony toothed jaws instead of the typical avian beak. The tail, too, retains bones—something that later birds would evolve to lose for reduced weight and more efficient flying. The wings are composed of full flight feathers, but have hooked claws still protruding prominently from the hands.

The asymmetry of the wing feathers shows they were evolved for aerodynamic performance and not just insulation or display, meaning *Archaeopteryx* was certainly using them for flight.

The fact that birds are descended from dinosaurs is now pretty common knowledge. However, it is often mistakenly thought that birds came after them, whereas, in fact, *Archaeopteryx* proves that dinosaurs were sharing the planet with birds for more than half of the Mesozoic era.

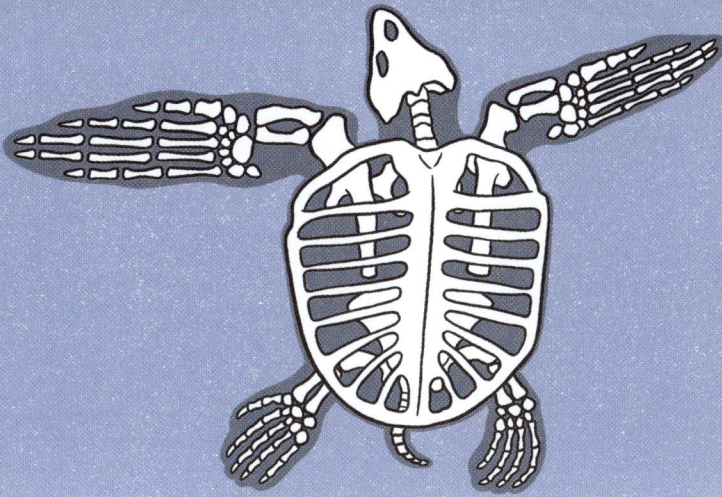

ARCHELON

AR-KE-LON

FIRST DISCOVERED: 1895 **RANGE:** USA

Turtles first appeared around the same time as the first dinosaurs, with the first full-shelled fossil turtle dated to the Late Triassic. Today, the leatherback turtle is the largest reptile in the sea, but even this giant was small compared to *Archelon*, the largest sea turtle in recorded history.

This behemoth of the Cretaceous period could reach 15 feet (4.6 m) in length, almost twice the size of what would be considered a large leatherback.

In appearance, *Archelon* was very similar to modern sea turtles. It had four scaley flippers, a beaked mouth, and a body-encasing shell (called a carapace). The structure of their jaws suggest they may have been specialized for eating tough invertebrate prey, likely spending much of its time settled on the sea floor feeding on mollusks. This is backed up by studies on its flipper anatomy, which suggest it was a weaker swimmer than modern open ocean-going turtles like loggerheads.

One theory identifies the warmer waters of the Cretaceous period as a possible factor in explaining *Archelon's* magnificent size. It's a theory potentially supported by the fact that the top three largest turtle species ever discovered were all dated to the last 15 million years of the Mesozoic.

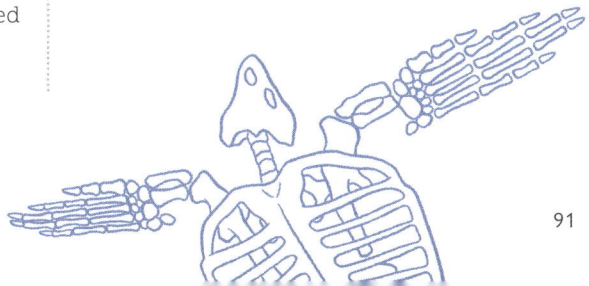

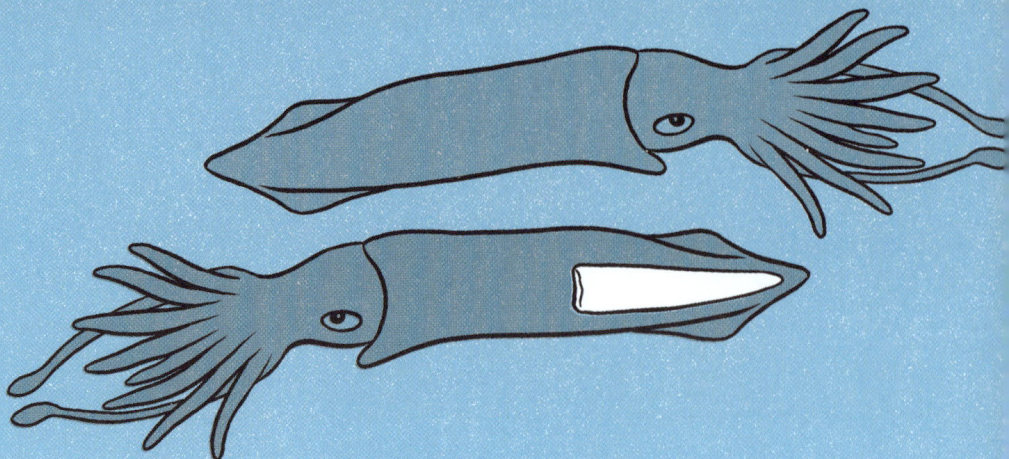

BELEMNITE

BELL-EM-NITE

FIRST DISCOVERED: UNKNOWN (NAMED IN 1546) **RANGE:** WORLDWIDE

Some of the most common marine fossils are the bullet-shaped belemnites. The conical fossil found today is the rostrum, located at the apex of the cephalopod's body, which otherwise greatly resembled a squid's. The presence of this structure, used for balancing the animal, also represents one of the largest differences between belemnites and squid.

We only know the rostrum's position, and the other features of the animal (like tentacles and "wing-fins"), thanks to exceptionally preserved specimens showing soft body features. Some even preserve ink sacs, which would have been used to evade predators. Mostly, though, we have only been left with the calcite-composed rostum (and occasionally another internal structure called a phragmocone, which is made of aragonite and also conical in shape).

The largest known belemnite, the genus *Megateuthis*, is estimated to have reached up to 10.2 feet (3.11 m) in length, which is comparable to the giant squids we know today.

First evolving in the Triassic, the belemnites went extinct with the dinosaurs at the end of the Cretaceous. Their fossils are often found in huge abundance, a testament to their success and possibly hinting at mass dying events after mating— again, just like modern squid.

CRINOID

CRY-NOID

FIRST DISCOVERED: UNKNOWN (NAMED IN 1847)   **RANGE:** WORLDWIDE

Crinoid fossils often appear as circular disks, with a hole in the center like a flattened donut. These structures are known individually as ossicles, and together they stack to create the stem of a crinoid, a common marine invertebrate of the ancient seas.

Sitting atop the stem is the calyx, the main body of the animal, from which the feeding fronds protrude. In appearance, the stem and crown of a crinoid look almost like a flower, giving them their common name of "sea lilies."

Extending up into the water, crinoids rely on filter-feeding particulate food out of the current. This is not an entirely passive process, as crinoids today have been observed maximizing their feeding efficiency by moving their arms to best suit the direction of the current.

Crinoids first appear in the fossil record well before the dinosaur age; the oldest confirmed crinoid was dated to the Ordovician period, about 480 million years ago. These highly successful invertebrates can still be found living in the seas, though they are much less diverse and common than at their peaks in the past. About 700 species are alive today, whereas there are over 5,000 known species that are now extinct.

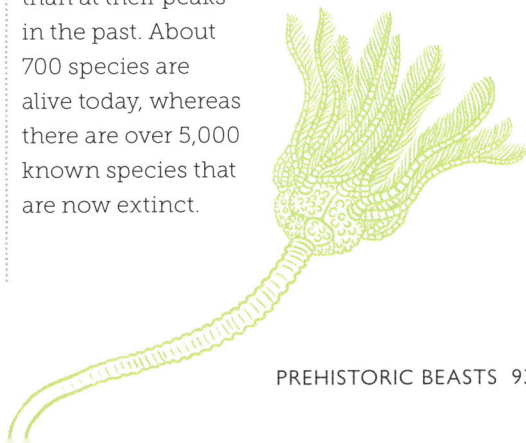

ENANTIORNITHES

EN-ANT-EE-OR-NEETHS

FIRST DISCOVERED: 1971 **RANGE:** WORLDWIDE

All of the birds alive today are part of the same taxonomic class: the Aves. During the Mesozoic era, however, there were other lineages diversifying in the hopes of dominating the skies. Although the Aves may have won this battle, for a time, another group, called the Enantiornithes, were real contenders.

The biggest difference between the Enantiornithes birds and true birds was that most Enantiornithes retained a tooth-filled jaw and external claws on their wings, clearly showing their dinosaur ancestry. Their jaws, however, were of a smaller size compared to their fully ground-dwelling dinosaur relatives, which reduced their weight.

The tails of Enantiornithes were unique, as they had neither the fan of feathers seen in modern birds, nor the long bony structures of dinosaurs. Instead, they had short tails, not used in flight at all, but with extended display feathers likely used to attract mates, like modern birds of paradise. Initially, Enantiornithes birds had a generalist diet, but as they evolved and diversified, they began to carve out ecological niches, each species specializing in a specific diet, from hard-shelled insects to soft fruits. Currently we know of over eighty species of Enantiornithes, all from the Cretaceous period. None of these survived past the dinosaur extinction, however, meaning this group of birds represents an evolutionary dead end on the Aves family tree.

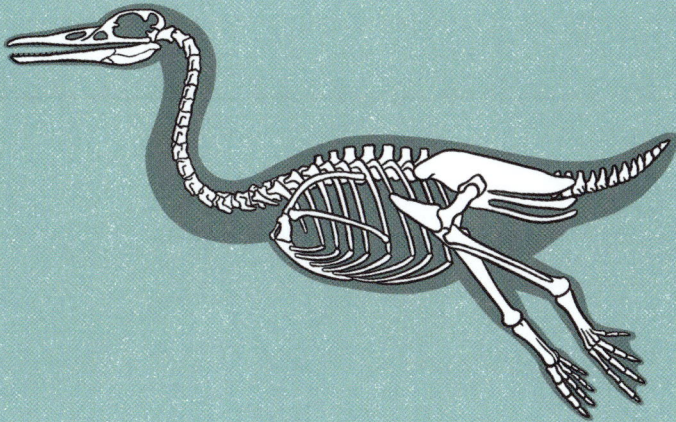

HESPERORNITHES

HES-PER-OR-NEETHS

FIRST DISCOVERED: 1888 **RANGE:** NORTHERN HEMISPHERE

In the modern world, penguins are the masters of diving birds, but they did not evolve until after the Mesozoic era. During the dinosaur age, the first birds to take to the water were the Hesperornithes.

These birds were notable for their extremely small wings, as well as huge, powerfully built legs that were positioned far back on the body. These, along with their webbed feet, allowed Hesperornithes birds to propel themselves through the water. Like penguins, Hesperornithes had beaks, but theirs were full of narrow teeth, perfect for seizing their fish prey.

Hesperornithes diverged from non-avian dinosaurs later than the Aves, and we have only discovered their remains dating from the Cretaceous period, meaning they lived around 145 to 66 million years ago. They were mostly confined to the North American continent, though one species has been described from Eurasia.

The largest known hesperornithe bird was *Canadaga*, an Arctic-dwelling swimmer which could potentially have reached over 7 feet (2 m) in length. Some studies have suggested that Hesperornithes species evolved to be larger at high latitudes, though with limited fossils to work with, this could be simply an artifact of bias in preservation.

ICHTHYOSAURS

IK-THEE-O-SAWS

FIRST DISCOVERED: UNKNOWN (EARLIEST RECORD FROM 1699) **RANGE:** WORLDWIDE

Before people knew about the dinosaurs, the ichthyosaurs were the celebrities of the fossil reptile world, even getting a mention in popular works of science fiction such as Jules Verne's *Journey to the Center of the Earth*. The name ichthyosaur means "fish-lizard," which is a fitting description of their body shape, with a dorsal fin and tail similar to that of a shark, but a very reptilian skull.

Ichthyosaurs first appear in the fossil record very early in the Mesozoic, around 248 million years ago—over 10 million years before the first dinosaurs.

The long snout (rostrum) of the ichthyosaurs was perfect for catching fish in the ancient seas, whereas some species, like *Temnodontosaurus*, were large and strong enough that they would have hunted fellow marine reptiles instead. Others, like *Excalibosaurus*, had an upper jaw that was far longer than its lower one, likely for sifting through the sediment to disturb small prey, which it would then snap up.

The common depiction of an ichthyosaur reflects those seen during the Jurassic, but the ichthyosaurs actually showed some of their greatest diversity during the Triassic. At the

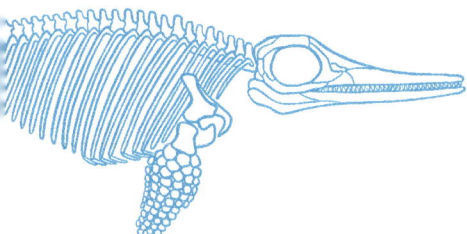

very end of this period is also when species reached their largest sizes, like the 65-foot-long (20 m) *Shastasaurus*. It's possible that they got even bigger than this; only known from fragments, *Ichthyotitan* from the UK may have been around 82 feet (25 m) in length, which would make it the largest known marine reptile ever.

Even more fascinating is the speed at which ichthyosaurs evolved to become so large. A discovery of a 6.5-foot-long (2 m) skull of a *Cymbospondylus* ichthyosaur from Nevada suggests they reached such gigantic sizes after just 8 million years, which is pretty rapid in evolutionary timescales.

Although all ichthyosaurs retained a vaguely similar shape overall, their bodies did vary between species depending on their lifestyles. For example, the active water hunters evolved further streamlining to reduce drag, whereas others had more widened paddle fins for slow patrolling of the sea floor.

Despite their early burst of rapid evolution, the ichthyosaurs could not follow that initial success all the way through the Mesozoic era, becoming extinct around 90 million years ago during the Cretaceous period—over 20 million years before the extinction event that would wipe out the dinosaurs.

COUNTER-SHADING

Many ichthyosaurs were dark on the top and light underneath—a type of camouflage known as counter-shading. When viewed from above, the animal would have been harder to see against the murky sea depths; when viewed from below, potential predators would have has a hard time seeing it against the sunlight. Counter-shading is common among marine animals today, including the great white shark.

LEEDSICHTHYS

LEE-DES-IK-THIS

FIRST DISCOVERED: 1886 **RANGE:** UK, FRANCE, AND CHILE

During the Late Jurassic, when the sauropod dinosaurs were among the largest animals to walk the Earth, the oceans were inhabited by *Leedsichthys*, one of the planet's largest known fish. This leviathan is thought to have reached over 52 feet (16 m) in length (roughly the same as a sperm whale), weighing up to 45 tons (45,722 kg).

Because the composition of many fish skeletons makes them less resilient to the fossilization process, there isn't much material to go on when it comes to *Leedsichthys* fossils. Study of the available bones of the skull and gill arch, however, suggests that

Leedsichthys was a suspension feeder, like a modern basking shark, filtering particulate food from the water.

Being one of the largest fish in the sea did not necessarily make *Leedsichthys* invulnerable, however; one specimen appears to show signs of bone regrowth, implying it may have survived an attack from a marine reptile in its lifetime.

Having been initially found near (and named for) the English city of Leeds, further finds in what is now Chile have suggested *Leedsichthys* was in fact a global ocean traveler.

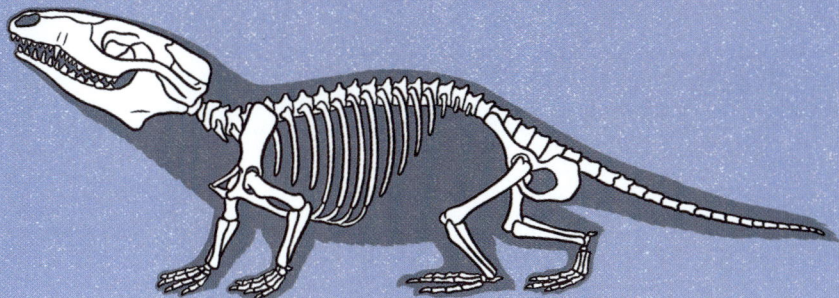

MORGANUCODON

MOR-GAN-UH-COE-DON

FIRST DISCOVERED: 1947 **RANGE:** NORTHERN HEMISPHERE

The earliest mammals living during the Mesozoic era could usually be described as small, insectivorous, and largely nocturnal. Though not a true mammal, all of these features can be seen in *Morganucodon*.

This small mammaliaform can be dated to the very end of the Triassic, about 205 million years ago. Known mostly from quarries in Wales, *Morganucodon* informed much of what we know about the time when many distinctive mammalian features were first evolving.

In modern mammals, the lower jaw comprises a single bone, the mandible. Other bones that were previously located in this region evolved to massively reduce in size and change position to become what now form our inner-ear bones. In *Morganucodon*, these bones are still present as part of the jaw, but show a huge reduction in their load-bearing function, which shows us that they were used primarily for hearing, not jaw articulation.

Analysis of *Morganucodon* teeth have shown that these creatures lived far longer than you would expect a similarly sized modern mammal to, suggesting their metabolism was potentially more similar to that of a reptile. However, looking at the nutrient passages within the bones shows that they were well on their way to evolving the warm-blooded metabolism seen in their descendants.

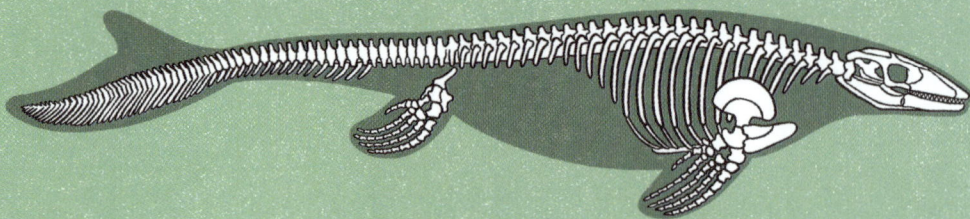

MOSASAURS

MOW-SA-SAWS

FIRST DISCOVERED: 1764 **RANGE:** WORLDWIDE

The mosasaurs were the top predators of the marine realm during the end of the Cretaceous. Although only around for the final 15 million years of the Mesozoic era, their fossil distribution shows that they were successful in colonizing much of what we now know as the Atlantic Ocean.

Superficially, mosasaurs share similar features with other marine reptiles. However, the way they move gives away their true familial line. Mosasaurs do not propel themselves with paddling flippers or a fish-like tail, but through an undulating motion like snakes, which are among their closest relatives.

Analysis of their teeth has revealed that mosasaurs were generalist predators, seemingly happy to feed on a range of prey, from fish and hard-shelled invertebrates to other large marine vertebrates.

In recent years, mosasaur popularity has surged thanks to their appearance in the *Jurassic World* franchise, but this comes with a swell of misconceptions. The real mosasaurs would not have been covered in thick crocodilian feature scales, but were likely smooth to aid in streamlining. They are also estimated to have topped out at 42 feet (13 m) in length, as opposed to the 100 feet (30 m) behemoth featured in the films.

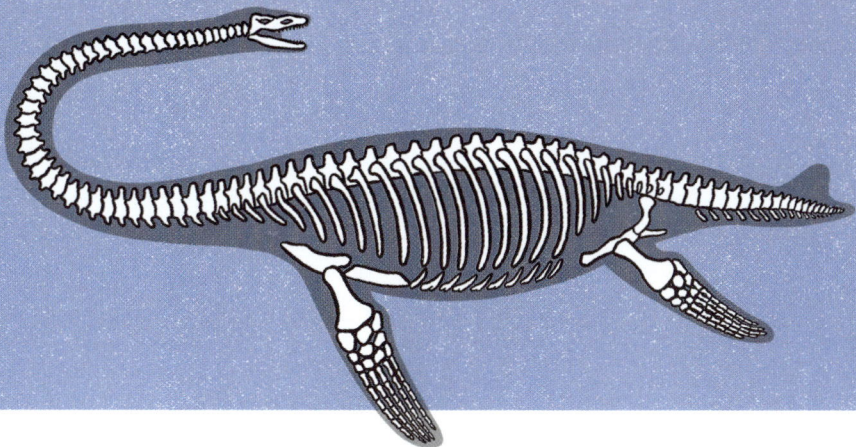

PLESIOSAURS

PLE-SEE-O-SAWS

FIRST DISCOVERED: 1605 **RANGE:** WORLDWIDE

Thanks to the story of the Loch Ness Monster, the body shape of a plesiosaur has become just as iconic in the world of mythology as it has in paleontology. The monster might not exist, but the animals were very real, and could be found swimming the seas for 140 million years.

Recognizable from their long necks, the plesiosaurs are thought to have used these to nimbly snatch fish from within shoals. Hydrodynamic tests of various plesiosaur body shapes have also shown that, when held in line with the body, a long neck does not noticeably increase the drag the animal experienced when swimming.

One Canadian plesiosaur, *Albertonectes*, measured 36 feet (11 m) in total length and 23 feet (7 m) of that alone was neck, containing over seventy vertebrae.

Plesiosaurs would swim using their four flippers, and had a specially reinforced chest, clearly visible in their fossils, which would have allowed them to make extremely powerful swimming strokes.

Although the plesiosaurs survived through more of the Mesozoic than the ichthyosaurs, just like the dinosaurs on land, these giant reptiles didn't make it through the end-Cretaceous extinction event that occurred around 66 million years ago.

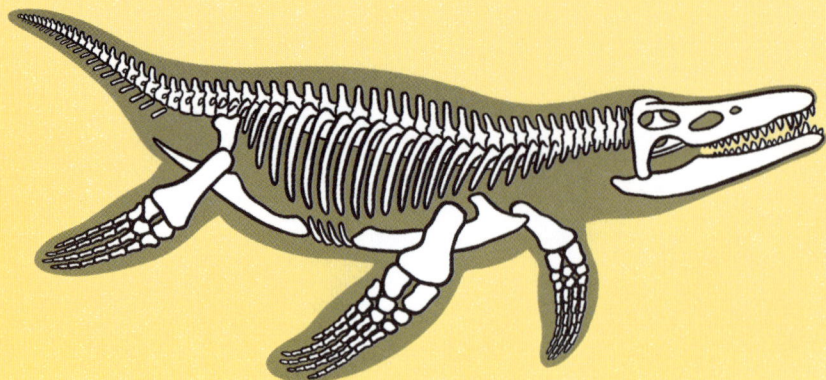

PLIOSAURS

PLY-O-SAWS

FIRST DISCOVERED: 1841 **RANGE:** WORLDWIDE

With mouths up to 6.5 feet (2 m) in length filled with 8-inch (20-cm) teeth, pliosaurs were among the most formidable-looking predators the world has ever known. These marine reptiles were related to plesiosaurs, moving through the water with four flippers and possessing specially reinforced ribs structures in their bellies.

The pliosaurs's necks were short, rather than long, and their jaws far more forceful. Whereas plesiosaurs evolved to hunt shoals of fish, pliosaurs were clearly specialized to hunt other large vertebrates.

Testing of these jaws has revealed them to be very powerful indeed, able to exert a force potentially twice that of a great white shark, and capable of catching and eating prey up to half their full size.

However, while a pliosaur's bite was strong, its jaws were surprisingly weak to rotational forces, meaning it wouldn't have thrashed its prey around like a crocodile. This is likely due to an evolutionary compromise, trading off some strength for a more streamlined shape.

The pliosaurs were among the most dominant marine predators throughout the Jurassic, though did not fair so well in the Cretaceous, and had likely died off long before the dinosaur extinction. The latest known fossil pliosaur, *Megacephalosaurus*, was found in the USA and dated to about 93 million years ago.

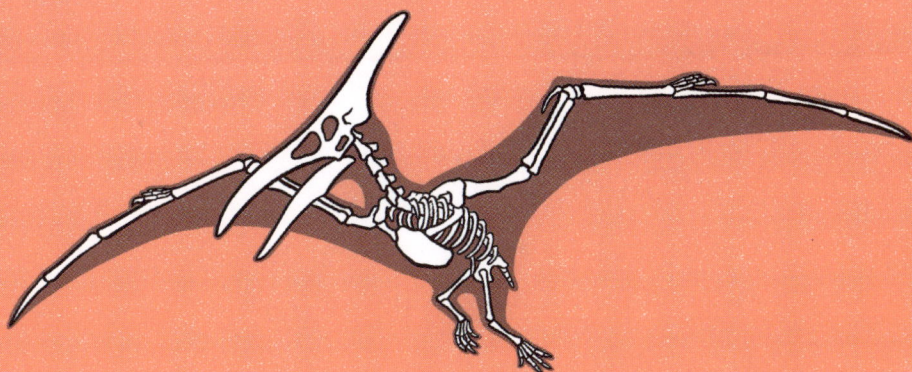

PTERANODON

TER-AN-O-DON

FIRST DISCOVERED: 1871 **RANGE:** NORTH AMERICA

One of the most iconic silhouettes of the prehistoric world is the winged and crested form of the *Pteranodon*. This pterosaur had a wingspan over twice that of the largest flying bird today (the wandering albatross), stretching out around 21 feet (6.5 m) from tip to tip.

It is believed that the primary function of the *Pteranodon* crest was to act as a display structure. It has been argued that differences in the shapes of the crest are indicative of different sexes; males are believed to have had the larger crests, whereas females displayed reduced structures.

It was long thought that *Pteranodon* was a fish-eater, thanks to fish fossils associated with *Pteranodon* remains, but whether it may have water-skimmed and eaten fish on the wing is still uncertain. The narrow, toothless beak could snap shut rapidly to effectively grab struggling prey. However, studies of the neck suggest it was too weak to withstand the forces created by entering the water at speed.

Fossils of *Pteranodon* have been discovered across North America, and it would have soared in the skies toward the end of the Cretaceous period, dying out around 84 million years ago. This means, contrary to popular depictions, that *Pteranodon* would not have been flying above the heads of *T. rex* and *Triceratops*, who wouldn't evolve for another 16 million years or so.

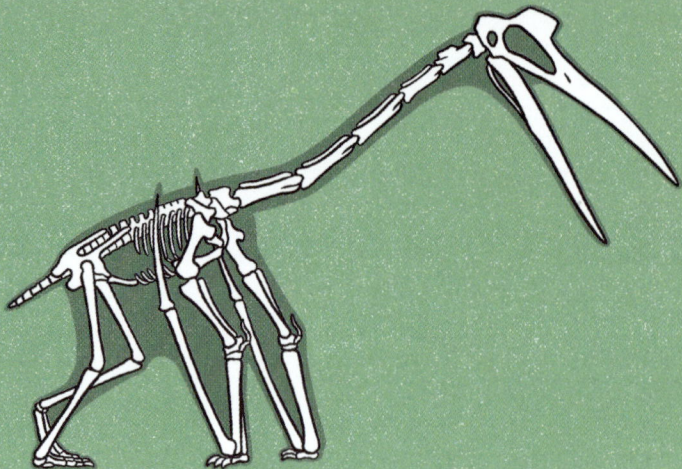

QUETZALCOATLUS

KET-ZAL-CO-AT-LUS

FIRST DISCOVERED: 1971 **RANGE:** NORTH AMERICA

With a wingspan comparable to that of a F-16 fighter jet (around 33 feet [10 m]), *Quetzalcoatlus* was one of the largest animals ever to fly. It was also among the very last pterosaurs, living through the final days of the Cretaceous in what is now North America.

The long neck and giant wings of the pterosaur inspired its name, which is taken from the Aztec god Quetzalcoatl, who takes the form of a flying serpent.

Despite being almost as tall as a giraffe when standing, *Quetzalcoatlus* was extremely light: the whole animal is estimated to have weighed only around 485 lbs (220 kg)—about the same as a donkey. Its bones were lightly built but still strong, anchoring the powerful flight muscles on the pterosaur's chest.

Although clearly built to fly, studies of their beaks have hinted that *Quetzalcoatlus* would have spent much of its time feeding on the ground. In this way they are often likened to modern storks, using their long necks and beaks to reach overhead and snap up prey close to the ground. For

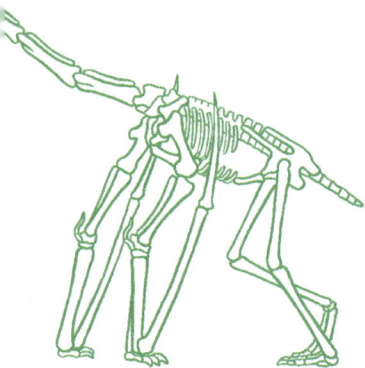

an animal as large as *Quetzalcoatlus*, young dinosaurs may well have been a regular food of choice.

Spending so much time on the ground means that *Quetzalcoatlus* must have had a very efficient system for getting itself into the air again. Research on the power they were capable of generating with different take-off techniques has shown that they would likely have done so quadrupedally (from all fours), jumping with their back legs and pushing off powerfully from their arms, before launching themselves up and forward to the point where the wings could open fully and flap to gain more lift. *Quetzalcoatlus* was not the only giant pterosaur, and the remains of members of its family, the Azhdarchidae, have been uncovered across the planet, including *Hatzegopteryx* in Europe and *Azhdarcho* in central Asia. Even though they are known from several species, pterosaurs remain among the most enigmatic of extinct animals. Their fragile bones are often destroyed rather than preserved in the fossilization process, and with no animal alive today sharing their body plan, it is exceptionally difficult to attempt reconstructions.

PTEROSAUR TAILS

Like all late-Mesozoic pterosaurs, *Quetzalcoatlus* had a highly reduced tail. However, earlier pterosaurs, like *Dimorphodon*, still had long bony tails. These tails may have helped them improve their aerodynamic maneuverability in the air, allowing them to quickly change direction. However, as pterosaurs became larger, the burden of carrying that extra weight exceeded any benefits.

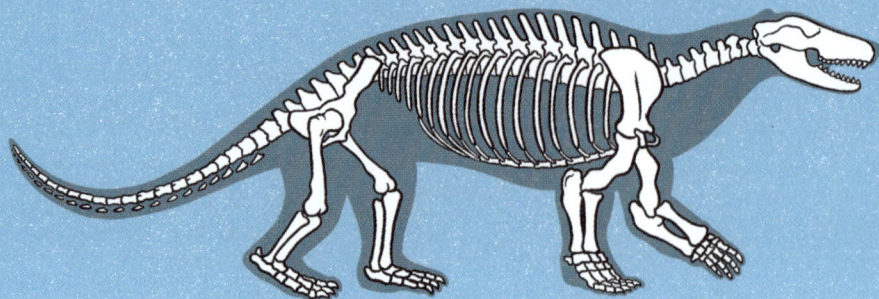

REPENOMAMUS

REP-EN-O-MA-MUS

FIRST DISCOVERED: 2000 **RANGE:** EASTERN CHINA

Despite evolving at a similar time to the first dinosaurs, the mammals never achieved the global dominance or massive sizes of their reptilian rivals. One of the largest mammals discovered from the Mesozoic was *Repenomamus*, around 3 feet (1 m) from head to tail and comparable to our modern-day badger.

Repenomamus lived during the Early Cretaceous, in what is now China. It was a primitive mammal, not within any of the groups we recognize today, and would still have been an egg-layer, as opposed to giving birth to live young.

In 2005, *Repenomamus* achieved infamy when a specimen was found that appeared to have been attacking a juvenile dinosaur (a *Psittacosaurus*). Though long assumed that some of these early mammals would have

diversified their diet to include vertebrates, this was the first direct evidence of a mammal preying on dinosaurs.

Repenomamus held the crown of the largest known mammal of the Mesozoic era until 2024, when *Patagomaia* was announced from the Late Cretaceous of South America. Although only known from fragments, *Patagomaia* has been estimated to have weighed as much as 31 lbs (14 kg)—about 4 lbs (2 kg) more than *Repenomamus*.

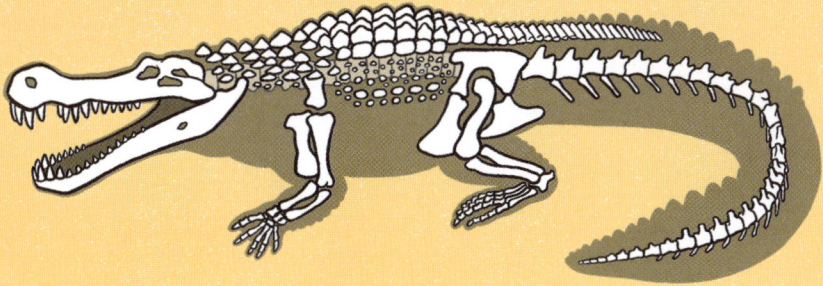

SARCOSUCHUS

SAR-COE-SU-KUS

FIRST DISCOVERED: 1867 **RANGE:** NORTH AFRICA AND BRAZIL

Though it's often said that crocodiles are older than the dinosaurs, this isn't exactly accurate. The group of crocodiles we know today actually only evolved toward the very end of the dinosaur age, whereas it was their superficially similar relatives that thrived throughout the Mesozoic. Some, like *Sarcosuchus*, became absolutely enormous.

At 4 tons (4,064 kg) and over 30 feet (9 m) in length, *Sarcosuchus* was around twice the size of the average saltwater crocodile (the largest reptile alive today). Large enough to take on dinosaurs drinking at the water's edge, *Sarcosuchus* would likely have relied on a quick snap to kill its prey, since the narrow snout likely wouldn't have been able to withstand the massive strains of the "death roll" employed by many modern crocodiles.

Sarcosuchus's body was reinforced with osetoderms (bone-deposits within the skin). Cross-sectional analysis of these showed that *Sarcosuchus* may have lived for over fifty years, though speedy growth as juveniles meant that they reached their full size well before then.

Sarcosuchus was a top predator in the waterways of the southern hemisphere during the Early Cretaceous, but it wasn't the only giant crocodilian of the Mesozoic. *Deinosuchus*, an alligator relative which lived later and farther north, may have weighed twice as much.

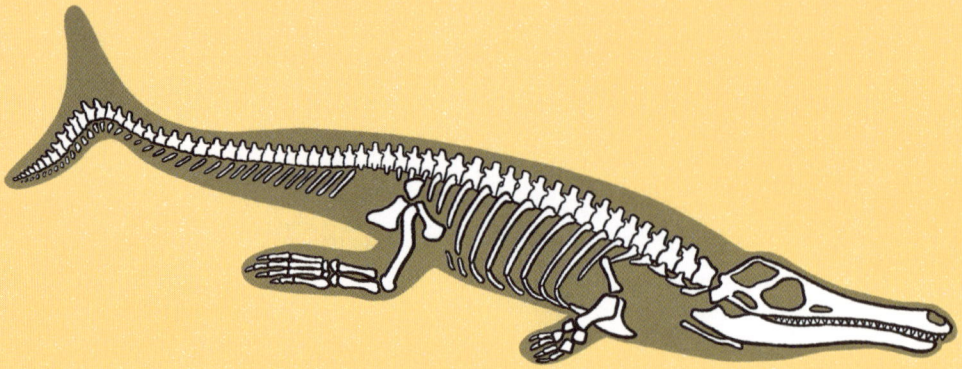

THALATTOSUCHIAN

THA-LAT-TOE-SU-KEY-AN

FIRST DISCOVERED: 1901 **RANGE:** WORLDWIDE

Modern crocodiles may spend a lot of time in the water, but they retain many of their terrestrial adaptations. Their Jurassic relatives the Thalattosuchians, however, included species adapted to live entirely aquatic lifestyles, showing just how diverse the crocodylomorph family tree once was.

There were two major groups of Thalattosuchians: the more familiar-looking Teleosauroidea (often said to resemble gharials) and the fully marine Metriorhynchoidea. Those which adapted to living fully in the open ocean had arms more like flippers and tails capped with fish-like flukes for more powerful propulsion.

Notable too was the texture of Metriorhynchoidea skin, which was smooth with no visible scales. They had traded the bulky armor we associate with modern crocodilians for a streamlined, fibrous texture, similar to that seen in marine reptiles like ichthyosaurs.

Studies on the hip structure of the Metrioryhnchoidea have also led to speculation that these animals may have given birth to live young rather than laying eggs. This is a rare trait among reptiles, but is seen in some other aquatic venturing species, such as modern water snakes.

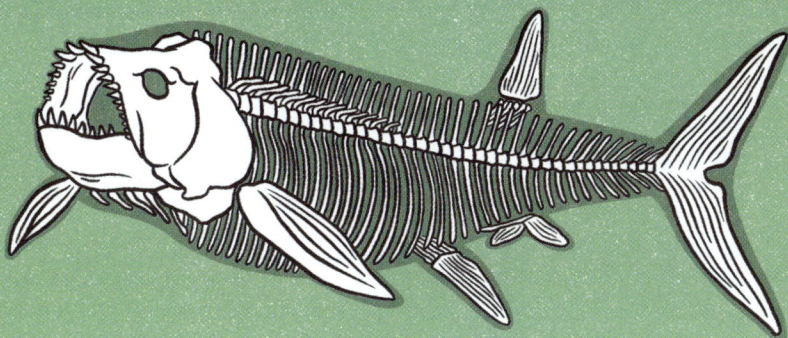

XIPHACTINUS

ZI-FAK-TIN-US

FIRST DISCOVERED: 1850s **RANGE:** WORLDWIDE

With a mouth full of vicious-looking teeth, the skeleton of *Xiphactinus* shows that the sharks were clearly not the only predatory fish in the oceans of the Mesozoic. The impressive nashers aside, it is the size of *Xiphactinus* that truly stands out. Sixteen feet (5 m) in length, this Cretaceous fish was even larger than the modern-day great white shark.

The sleek body plan of *Xiphactinus* suggests it was an agile hunter, able to speed through the water to catch other fast-swimming fish. We know that it would tackle large prey too, thanks to an extraordinary fossil that shows an entire 6.5-foot-long (2 m) fish within the stomach of a *Xiphactinus* specimen.

Rare specimens of *Xiphactinus* juveniles reveal that, like many modern large marine animals, they spent their younger years in the shelter of shallow coastal regions before moving out to the open seas once they'd grown. Their teeth are just as sharp in the young as they are in the adults, which implies they were predatory from the moment they hatched from their eggs.

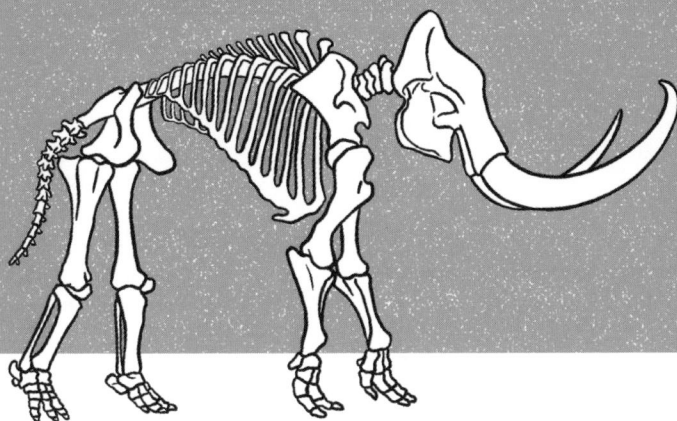

CHAPTER 3

EVENTS AND TIME

EVENTS AND TIME

Human civilization has existed on this planet for about 10,000 years, an enormous span that is reduced to nothing when compared to the vastness of geological deep time. Paleontologists studying the history of all life on Earth are dealing with a stretch of approximately 3.8 billion years. When faced with such an unfathomable concept, it only makes sense to split it into more manageable chunks, definitive units which are known collectively as the geological timescale.

This chapter will explore the geological timescale in chronological order, focusing primarily on the time during which the dinosaurs could be found living on Earth. It will also put this period into its greater context, examining the key events through life history that sculpted the world before, during, and after the dinosaurs.

The largest units of geological time are the "eons," which stretch hundreds of millions of years. Since the formation of the Earth 4.6 billion years ago, there have been four recognizable eons: the Hadean, Archean, Proterozoic, and Phanerozoic.

Eons are further divided up into "eras," which themselves are made up of geological "periods." The best known of these are the Triassic, Jurassic, and Cretaceous periods, which together make up the Mesozoic era, commonly referred to as the "Age of Dinosaurs." This is the era which makes up the focus of the majority of this book.

There are further divisions within

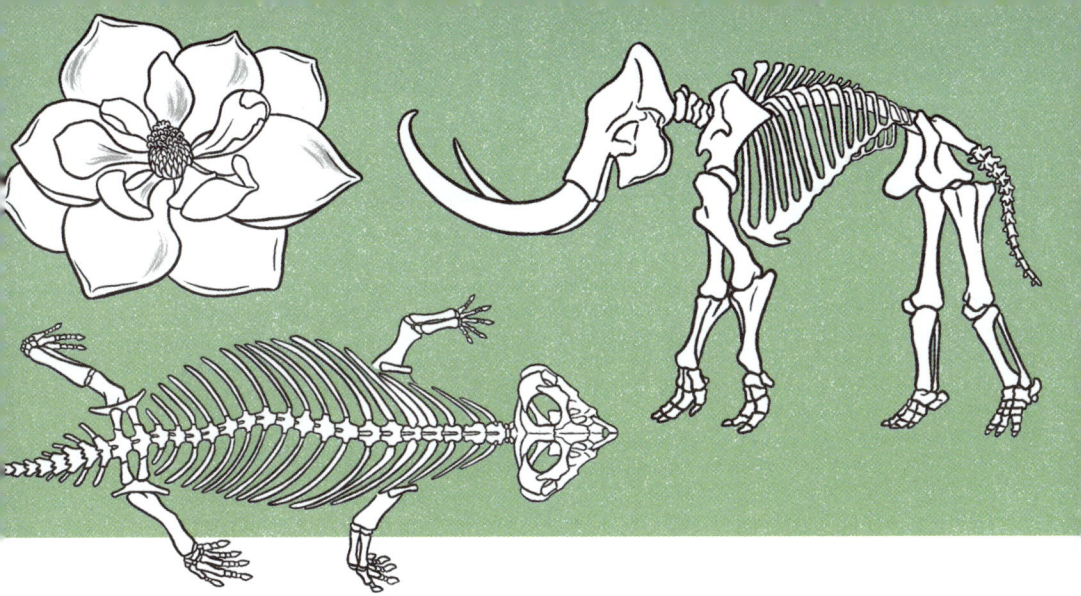

the eras, known as "epochs." During the Mesozoic era, these epochs are relatively simple, with the labels "Early," "Middle," and "Late" indicating when the epochs occurred within the era. The "Early" epochs are the oldest, at the start of each period, with the "Late" epochs being the more recent, found at the end of the period. Both the Triassic and Jurassic have Early, Middle, and Late epochs, whereas the Cretaceous has no formally defined Middle epoch.

Beyond just the Mesozoic, this chapter will discuss what paleontologists frequently refer to as the "Big Five" mass extinction events: times when the fossil record shows gigantic losses in the diversity of life over just a few million or even thousands of years. Such windows of time may seem vast on a human scale, but geologically they are genuinely rapid events, often marked in the fossil record by no more than a few millimeters of rock.

Once the dinosaurs were gone, the planet was by no means ready for us humans to immediately step into place, either. There were millions of years and many more key events that took place first, some of which are also touched on in this chapter.

In their 160 million years or so on the planet, the dinosaurs saw the continents moving, temperatures fluctuating, and sea levels rising and falling, meaning each stage of their evolution involved facing different challenges. When put together in sequence, they tell a truly fascinating story.

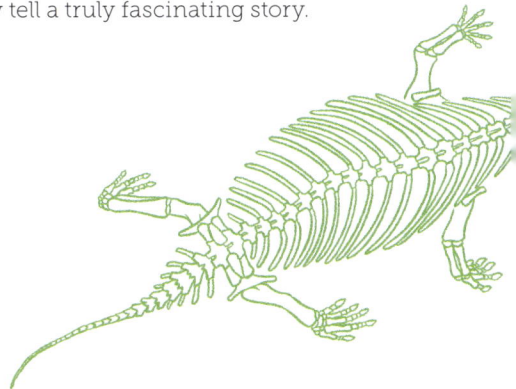

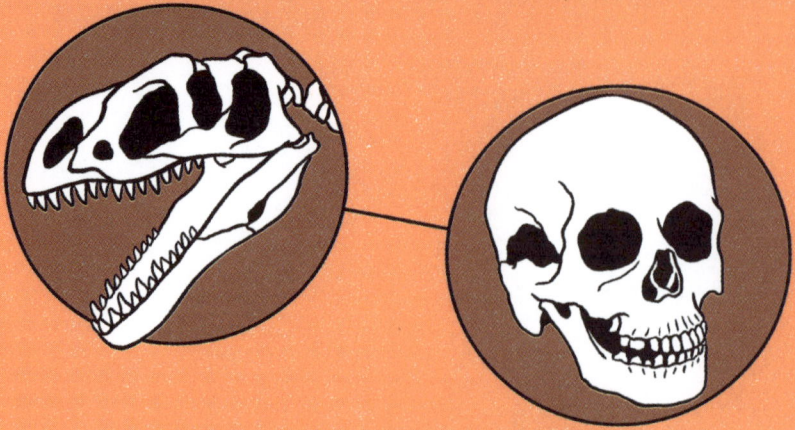

PHANEROZOIC EON

The Phanerozoic is the "Eon of Visible Life," and the current geological eon. Spanning back 538.8 million years, this eon contains not only us humans, but the entirety of the dinosaur reign, as well as the 300 million years before them.

The boundary marking the beginning of this eon was originally defined in the early twentieth century by the point in the rock at which fossils of complex life forms could first be found. In the century since this naming, we have found bountiful complex fossils from well before this point, making the boundary seem a little less clear.

Nevertheless, it is around this time that we see a shift from mysterious organisms with fairly sedentary lifestyles to those more interactive with the world around them. So, organisms started burrowing into the sediment rather than just lying atop it, swimming rather than merely floating, and actively pursuing their food, becoming true hunters.

To give an idea of how different the Earth is now compared to the start of the Phanerozoic, during the first period of this eon (the Cambrian), a day was only 20.5 hours long, as the speed of the Earth's rotation has slowed over time.

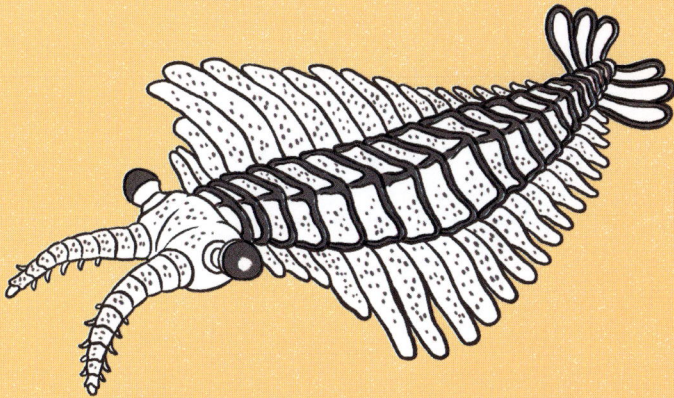

THE CAMBRIAN EXPLOSION

Though not a literal explosion, the Cambrian Explosion was no less dramatic. This milestone in life history occurred 538 million years ago and refers to a massive diversification of life, with many new groups of organisms appearing in the fossil record in a relatively short amount of time.

One of the biggest innovations driving this intense period of evolution was a hardened exoskeleton. Complex eyes also evolved around this time, as seen in trilobites, who had eye lenses made of crystalline minerals.

Shelled brachiopods and bivalves (similar to modern sea shells) and the arthropods (the group which would later contain insects and crustaceans) were among the well-known animals starring in this event. However, some are truly alien in appearance, like *Hallucigenia*, a half-spiked, half-tube-armed organism so bizarre that it took many years to even establish which side of the body was top and which was bottom.

Perhaps the most iconic creatures of the Cambrian Explosion were the anomalocarids. These predators swam through the water with undulating armored wings along their sides, found prey using their compound eyes, and caught them with spiked arms attached near their mouth.

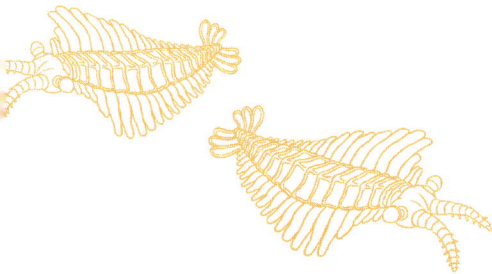

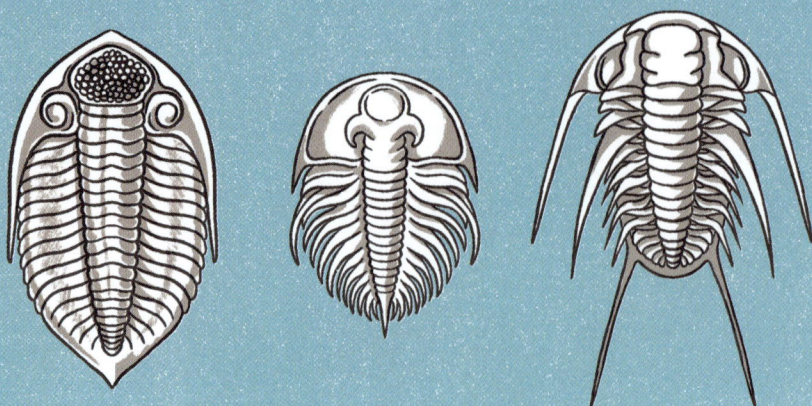

PALEOZOIC ERA

The first of the three eras within the Phanerozoic eon is the Paleozoic. Though not at all an exact measurement, this era could be thought of as "the era before the dinosaurs." It runs from the start of the Cambrian period 538 million years ago to the end of the Permian period 251 million years ago.

There may not have been dinosaurs yet, but the Paleozoic wasn't short of iconic animals. The trilobites, arthropods which superficially resembled woodlice, roamed the seas in their millions, growing up to 3 feet (1 m) in length.

The Cambrian period was followed by the Ordovician (485–444 million years ago), where giant cone-shelled squid relatives called orthocones could reach several meters in length. In the Silurian period (444–419 million years ago), the shallow coasts were patrolled by armored sea scorpions.

In the Devonian (419–358 million years ago), the armored placoderm fish evolved a huge diversity of head ornamentation. Most famously, the 11-foot-long (3.5 m) predator *Dunkleosteus*, whose jaws were formed of bladed bone, may have had one of the strongest bites of any fish.

By the Carboniferous (358–298 million years ago), life was well and truly established on land, in an age that saw the planet covered in dense swamp forests, and the rise of reptiles. Nevertheless, it is the magnitude of certain arthropods that draws the most attention, like the hawk-sized dragonfly *Meganeura* and the 8-foot-long (2.5 m) millipede *Arthropleura*.

The final period of the Paleozoic was the Permian (298–251 million years ago), which featured the iconic *Dimetrodon*. This sail-backed quadruped is often mistakenly labeled as a dinosaur, despite being separated from them by over 40 million years.

Also well known from this time were the saber-toothed gorgonopsians. These large predators were part of a group of animals known as therapsids, which we mammals are also a part of. However, like dinosaurs, mammals themselves did not evolve within the Paleozoic era, and animals like the gorgonopsians are remnants of a diverse early family tree which is now lost to time without direct descendants.

Just as the Paleozoic began with a major life event in the Cambrian Explosion, so too did it end. Only this time, it was not a mass expansion of life, but a loss: the Permian–Triassic Extinction.

SAIL-BACKS

Dimetrodon is the most iconic vertebrate animal of the Paleozoic era, but it was not the only sail-back of the time. *Edaphosaurus* also had large sails and looked similar enough to *Dimetrodon* that they were once grouped together as the pelycosaurs. However, we now know they were not that closely related, and the two evolved these sails independently.

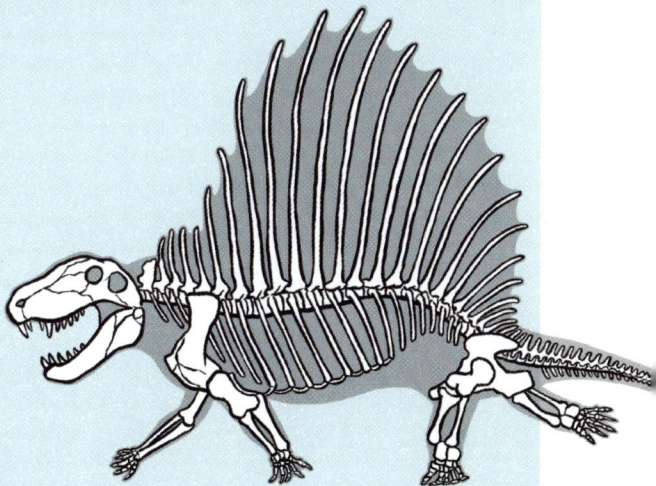

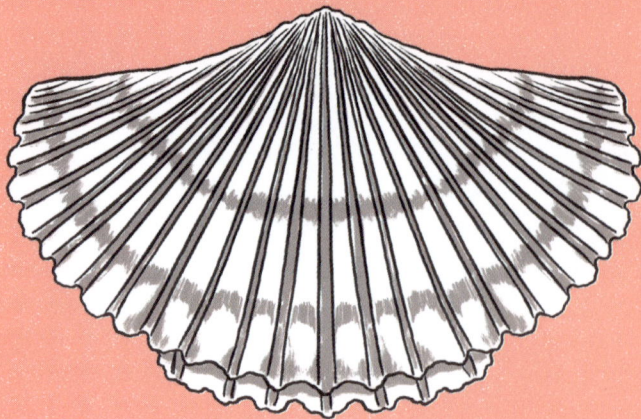

ORDOVICIAN–SILURIAN EXTINCTION

The first of the famed "Big Five" mass extinctions faced by complex life took place 444 million years ago, marking the end of the Ordovician period and the start of the Silurian. The extinction was caused by extreme climate change, when global temperatures and sea levels dropped substantially over just a few thousand years.

What exactly triggered these changes isn't entirely understood, but evidence for the effects is abundant. The southern continents show huge glacial deposits of a major ice age that would have drained the seas of billions of tons of water. Even where the water levels were unaffected, the changing climate caused large-scale lack of oxygen in the oceans—a disaster for marine life.

Unsurprisingly, the organisms most affected were the marine invertebrates, such as the brachiopods, though no major groups were lost completely. Rather than a complete overhaul of the dominant animals on the planet, we can think of this extinction as a severe setback in diversification. It's estimated that one hundred families of invertebrates faced extinction during this time, and it would take 5 million years for the Earth to recover its previous levels of diversity.

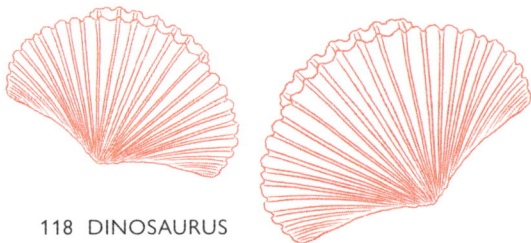

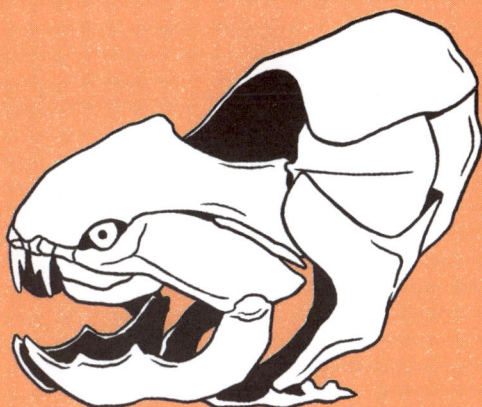

DEVONIAN
EXTINCTION

The second of the "Big Five" extinctions is a bit of a misnomer, as the Devonian Extinction occurred in two indistinct pulses, a few million years apart. This is perhaps the most gradual of the five, with widespread ocean anoxia causing havoc in the Earth's marine ecosystems. As with the first mass extinction event, it was the marine invertebrates that were the most affected.

The most prominent vertebrate victims of these extinctions were the placoderms, a group of armored fish with primitive jaws. This included the infamous predator *Dunkleosteus*, with bladed jaws capable of crunching through other placoderms.

The immediate aftermath of the extinction saw a decrease in body sizes of vertebrates across the globe, as the small, rapid-breeding animals were better able to cope in the challenging conditions. It would be many millions of years before they would once again match their previous sizes, and go on to far surpass them.

Thankfully for our own deep evolutionary line, most vertebrates weren't as strongly impacted by the Devonian Extinction. This is especially important considering what a significant time this was for the vertebrates, only a few million years after the first had emerged to colonize the land.

PERMIAN–TRIASSIC (P–T) EXTINCTION EVENT

The event that killed the dinosaurs may be the best-known extinction in Earth's history, but it was by no means the largest. In fact, the closest that complex life has come to being entirely wiped from the face of the planet occurred 250 million years ago, at the division of the Permian and Triassic periods (P–T), and it is an event so cataclysmic that it is regularly referred to as the "Great Dying."

The trigger for the event is thought to be a surge of volcanic activity that occurred to the north of the supercontinent Pangaea. This was the largest volcanic event of the entire Phanerozoic Eon, and saw enough erupted lava to cover a 3 million-square-mile (7.8m sq km) area in a few hundred thousand years. The remains of this eruption can be seen today across Russia and Kazakhstan, in a geological province called the Siberian Traps.

The amount of gases (such as carbon dioxide) released by the eruptions was enough to severely alter the global climate. Global temperatures rose and, as the seas were starved of oxygen, they became dangerously acidic to the shell-producing organisms living in them. As a result, ecosystems imploded the world over.

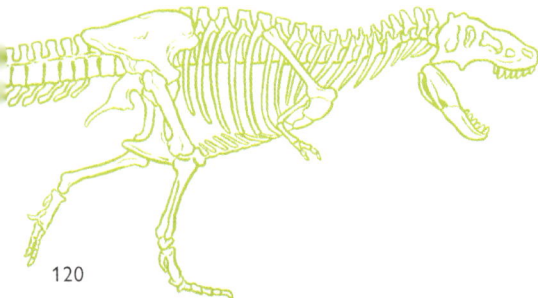

To make matters worse, much of life was still recovering from two smaller-scale extinction events that had occurred a few million years earlier. Some estimates suggest that species-level loss may have been above 90 percent across the planet.

Among the victims of this extinction were some iconic groups of prehistoric taxa, such as the trilobites. These three-lobed marine arthropods (which resembled, but were not closely related to, woodlice) had been thriving in the oceans since the Cambrian, over 270 million years previously.

Unlike the previous two of the "Big Five," the P–T extinction severely hit terrestrial life as well. The dominant vertebrates of the time were the synapsids, animals which in some ways resemble reptiles, but were in fact their own unique group.

Top of the food chain were the saber-toothed gorgonopsids, though even they were not resilient enough to survive the extinction. The synapsids that did survive included the smaller and adaptable cynodonts—a group that would later go on to evolve into the first mammals.

SURVIVORS

To survive the P–T, it helped to be a generalist, preferably small in body size, and able to adapt to challenges. All of these traits are personified by *Lystrosaurus*, perhaps the poster child of how to survive a mass extinction. These pig-tusked therapsids were so resilient that, in the immediate aftermath, they account for over 90 percent of all vertebrate fossils found in some regions.

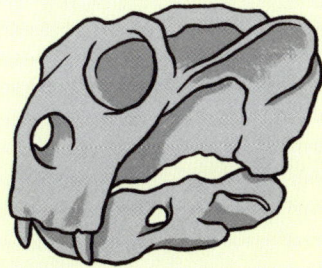

TRIASSIC PERIOD

The first period of the Mesozoic era was the Triassic (252–201 million years ago). The beginning of the period—the Early Triassic—is associated with the slow recovery of life from the P–T mass extinction, before leading in to the evolution of the dinosaurs, about 233 million years ago.

This means that, although associated strongly with the dinosaurs, they were not present for around the first third of the Triassic period. Instead, the early stages of the Triassic were dominated by "disaster taxa"—animals that thrived in the aftermath of the extinction. The tusked quadruped *Lystrosaurus* was particularly common, and could be found across all of Pangaea.

During the Triassic, the climate was warmer and generally more arid than it is on average today. Much of the heart of Pangaea would have been dominated by deserts and sparse scrub forests. However, monsoons would have allowed for denser and lusher foliage around much of the coasts.

Those forests would not have been as diverse as those we see today, and would have comprised conifers and tree ferns. Familiar floral groups like flowering plants and grasses were still many millions of years away from evolving.

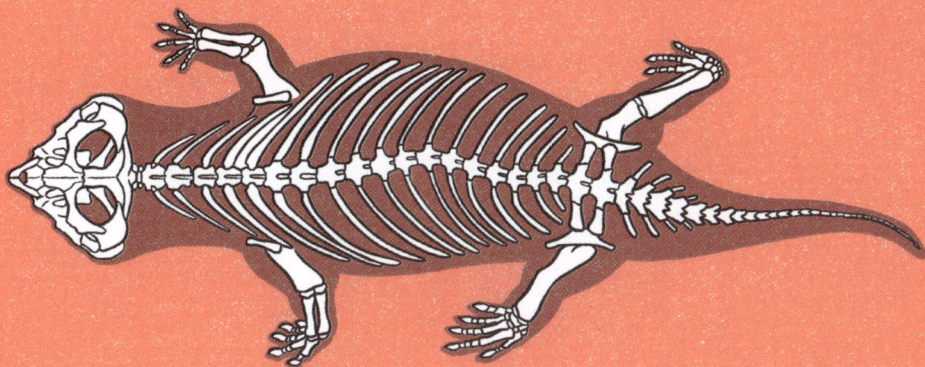

CARNIAN PLUVIAL EVENT

About 234 million years ago, part-way through the Triassic period, the rock record shows the world went through a major climatic shift. Defined by a huge increase in global rainfall and named for the geological stage of the Triassic in which it occurred, the Carnian Pluvial Event would have staggering implications for the future diversity of life.

Climate change, possibly driven by volcanism, saw a massive increase in humidity globally. For many groups, the change was a disaster and led to widespread extinctions (especially prevalent in the spiral-shelled ammonoids). But for others, it proved an opportunity.

The earliest Mammaliaformes (not yet true mammals) were among them, as were the ancestors of modern snakes and lizards. However, one particular group of reptiles really seemed to benefit from the ecological fallout, for it is within this episode of history that the fossil record first shows the appearance of the dinosaurs.

In some parts of the planet the fossils show them rising to dominance rapidly, but it would be some time before they achieved this globally. However, there can be no doubt that the Carnian Pluvial Event both drove and opened the door for the later success of the dinosaurs.

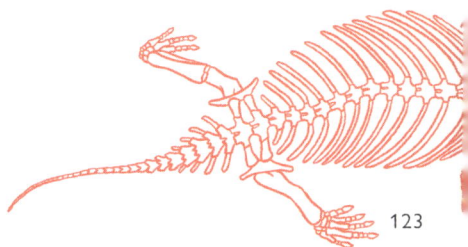

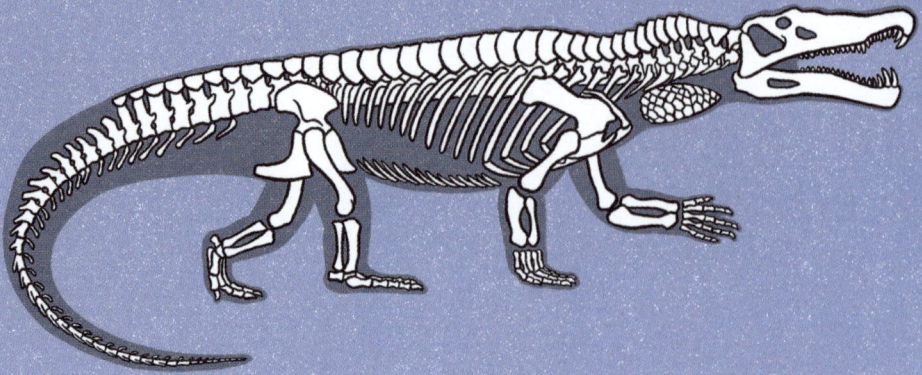

TRIASSIC–JURASSIC (T–J) EXTINCTION EVENT

The first major challenge faced by the dinosaurs was the third of the "Big Five" extinctions. Occurring 201 million years ago, this extinction event marks the boundary between the Triassic and Jurassic periods, which it is named for.

At this time Pangaea was beginning to break up, with a massive increase in volcanism at its heart. This rupturing event would go on to form the basis of the Atlantic Ocean, effectively ripping the continents apart. As always with such extensive volcanism, it was the resulting climate change that drove the extinction.

The dinosaurs would persist through this event, but many of their closest relatives would not be so fortunate. Dinosaurs are part of the archosaurs, today represented by just birds and crocodilians, but they once came in many more shapes and sizes. It is the fault of the T–J extinction that we don't now have galloping herbivorous crocodiles.

In the sea, the biggest loss were the conodonts. Though not widely known outside of paleontology circles, these small, jawless vertebrates had been present since the Cambrian, and their fossils have been extremely useful in dating particular units of time.

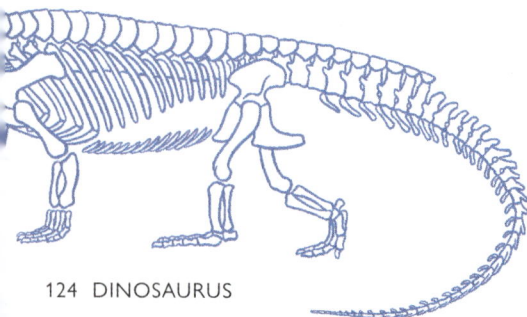

JURASSIC PERIOD

The middle period of the Mesozoic era is the Jurassic (201–145 million years ago), the time most associated with the dinosaurs. The climate was warmer on average than today, and there were no ice caps at the poles. Much of the world was lush and tropical, though areas of mass aridity did still exist, largely toward the eastern ends of the rupturing supercontinent of Pangaea.

It was during the Jurassic that dinosaurs became the dominant group of vertebrates on Earth, inhabiting every part of the planet and first reaching the gigantic sizes we associate with them.

A time of many important landmarks in Earth's story, it was also during the Jurassic that the first true mammals evolved from their furred, burrowing ancestors.

Unlike the Triassic, the end of the Jurassic is not marked by any major extinction event to divide it from the Cretaceous. Instead, there is a turnover of many animal groups, notably ammonoids, where those found before the boundary are significantly different than those found after it. However, this is not enough to have prevented the Jurassic–Cretaceous boundary from being described by some as perhaps the loosest in all of the Phanerozoic.

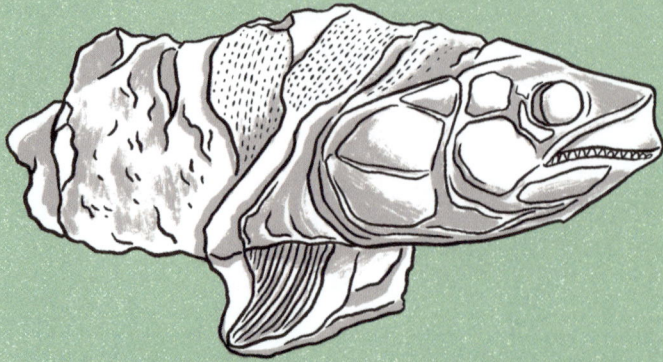

TOARCIAN OCEANIC ANOXIC EVENT (T-OAE)

Even though they do not breathe it like us, animals in the ocean still rely on oxygen to survive. Dissolved in the water, the animals can extract it with structures like gills. Sometimes, environmental factors can cause a dramatic reduction in the available oxygen, referred to as an anoxic event. The most devastating of these during the dinosaur age occurred during a time in the Jurassic known as the Toarcian, the T-OAE.

About 183 million years ago, a time of particularly active volcanism in the southern hemisphere caused a rapid increase in global temperatures and a rise in ocean acidity, resulting in widespread anoxic conditions. It was a disaster for marine invertebrates around the sea floor, who struggled to respire and couldn't form their shells due to the highly acidic waters.

The resulting domino effect caused localized marine ecosystem collapses. On land, the volcanic climate changes also spelled the end for several groups of dinosaurs, including *Dilophosaurus*, though the diversification that followed saw the arrival of several major dinosaur groups, such as the allosaurs and tyrannosaurs.

The T-OAE may have lasted for over half a million years, and it likely took several million more for Earth to recover the animal diversity it lost.

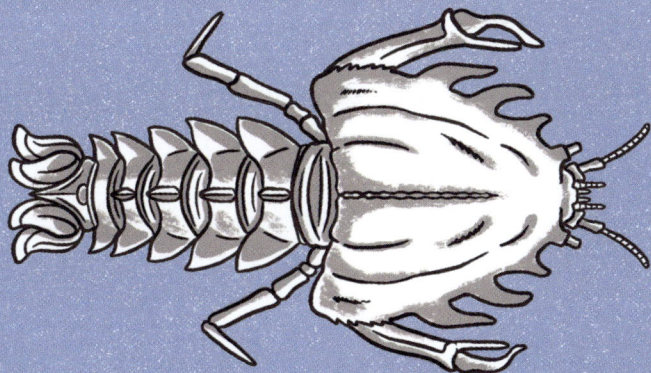

MARINE MESOZOIC REVOLUTION

Throughout the Mesozoic, there was an evolutionary arms race happening in the marine realm between predators and their hard-shelled prey. The resulting major diversification of species was known as the Marine Mesozoic Revolution (MMR).

As the supercontinent Pangaea broke apart, there were now more shallow marine environments on continental shelves than there had been previously, and many were more interconnected, meaning organisms were now in more direct competition with each other. As a result, new and novel predators began to evolve to exploit the abundance of food in these realms.

Many marine predators began to adapt specialized tools to crush the hard shells of invertebrates, a feeding style known as durophagy. Some common adaptations for doing this include rounded teeth and crushing beaks.

With predators now able to really exploit this source of prey, the result was a loss of diversity of certain invertebrates. Animals like sessile crinoids, anchored to the sea floor, never recovered their previous abundance.

It is believed that this event also influenced the evolution of true crabs, which did not appear in the fossil record until the Jurassic period.

CRETACEOUS PERIOD

The final period of the Mesozoic era was the Cretaceous (145–66 million years ago), the last geological period inhabited by the dinosaurs. Although it is regularly associated with their extinction, it was during the Cretaceous that dinosaurs reached their peak diversity. Some of the most famous dinosaurs, *Tyrannosaurus* and *Triceratops*, evolved during this period, and at no other time were they more successful and dominant.

By the Cretaceous, the world map was beginning to look more recognizable to us today. Pangaea had long since split up, though sea levels were still substantially higher (meaning there were many inland seas), and the Indian subcontinent was adrift and isolated in the ocean, yet to collide with Asia.

The climate, too, was more familiar, as the Cretaceous was the coolest period of the Mesozoic (both the Triassic and Jurassic being significantly warmer than today). However, there were still no permanent ice caps at the poles.

The name of the Cretaceous comes from the Latin name for chalk, which is a dominant rock type associated with this time period, especially across Europe where the periods were first defined.

ANGIOSPERM TERRESTRIAL REVOLUTION

Despite being so ubiquitous today (making up 90 percent of all plant species), flowers evolved more recently than most realize. The earliest fossil evidence of flowers is from 125 million years ago, during the Cretaceous. However, it is likely they first began to emerge in the Jurassic period.

When the flowering plants (known as angiosperms scientifically) did appear, they flourished, spreading across continents and altering the ecosystem of the entire planet. This remarkable event is known as the Angiosperm Terrestrial Revolution.

The success of angiosperms came largely thanks to the symbiotic relationship they developed with their insect pollinators, relying on each other to survive and breed. Many of the modern insects we are so familiar with, like bees and butterflies, got their starts during this revolution in Earth's history.

Although not part of the same revolution, another major plant group which appeared during the Cretaceous was Poaceae, the family we know simply as the grasses. As with the angiosperms, grass would not reach its current abundance until after the Mesozoic era, but it was under dinosaur feet that it began.

CRETACEOUS–PALEOGENE (K–PG) EXTINCTION EVENT

The most famous extinction event ever endured by life on the planet was that which occurred 66 million years ago, ending the reign of the dinosaurs. The event marked the division between the Cretaceous period ("K" in geological notation) and the Paleogene epoch ("Pg"), which gives the extinction its name.

Unlike other extinctions, which were triggered by drawn-out climactic change over millennia, the primary cause of the K–Pg occurred in a single instant, when an asteroid about 6.5 miles (10 km) wide impacted the Earth.

Remnants of the asteroid strike can be found across the world. When it vaporized on impact, a massive amount of the rare element iridium entered the atmosphere. As it settled out, it created a distinct layer which can be found at the end of the Cretaceous boundary the world over. Closer to the impact site are glassy spherules, remnants of the molten particulates thrown out as debris.

The immediate blast of the impact destroyed all life within miles, and triggered mega-tsunamis which wiped out ecosystems across the ocean. But the real killer was the ash and debris ejected into the atmosphere, which blocked out the sun for years.

This "long asteroid winter" would have thrown ecosystems into meltdown. Plants would have been unable to get the sunlight they needed to photosynthesize, and would die back in huge numbers. A dearth of food would have been a disaster for the large herbivorous dinosaurs, who needed huge quantities of foliage each day, quickly tumbling to extinction.

At first, opportunistic scavengers may have gotten by well enough, feeding on the remains of the dead. But it wouldn't be long before these too would have struggled to find food.

Size appears to have been a huge factor in what became extinct as a result of the impact. Though it's easy to think of them as the most resilient due to their size, the larger an animal is, the more vulnerable it can be to environmental changes. This was bad news for the dinosaurs, who were the most infamous victims of the catastrophe.

The key to survival was to be small, adaptable, and opportunistic. Those are three words that could certainly be applied to the mammals and the only dinosaurs to survive into the next era: the birds.

IMPACT SITE

The asteroid struck the Earth off the coast of the Yucatán Peninsula (highlighted on the map by the rectangular box) in Mexico (see Chicxulub Crater, page 147). By striking the water, the asteroid sent out a shock wave that generated tsunamis potentially 300 feet (100 m) in height. The sheer amount of energy created in that moment would have taken months to fully dissipate.

CENOZOIC ERA

The current geological era, stretching back from the human world to the end of the dinosaur age 66 million years ago, is the Cenozoic. Sometimes known as the time of the mammals, this is when these furry, milk-producing vertebrates diversified to take over every corner of the planet.

There are three periods that make up the Cenozoic: the Paleogene (66–23 million years ago); the Neogene (23–2.6 million years ago); and the Quaternary, our current geological period. The Quaternary is defined as the age of humanity, starting from when recognizable members of our species, *Homo sapiens*, first arose.

In the Cenozoic, mammals followed birds into the skies with the diversification of bats. First appearing in the fossil record 55 million years ago, there are now around 1,300 species of bat, more than any other mammalian group.

Around 50 million years ago, a group of ungulates (hooved mammals— including hippopotamuses) began exploring opportunities for living in close association with the water. Eventually evolving to be fully aquatic, descendants of these pioneers would become the largest animals ever to live on the planet: the whales.

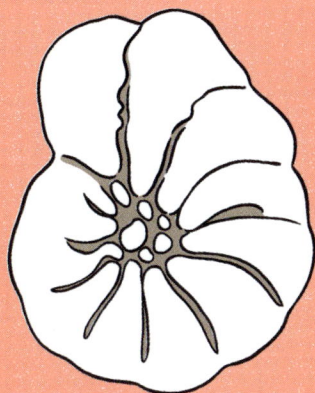

PALEOCENE–EOCENE THERMAL MAXIMUM (PETM)

Perhaps the most important prehistoric event that few have ever heard of is the Paleocene–Eocene Thermal Maximum (PETM). This period of swollen global temperatures occurred about 56 million years ago, and is the focus of intense study today as possibly an analog for understanding how the world may respond to contemporary climate change.

For about 200,000 years, global temperatures were in the region of five degrees warmer than the average. As with a number of the other global climate excursions, this one is believed to have been triggered by a period of volcanic eruptions.

Among the impacts of the PETM was the one positively identified ocean anoxia event of the Cenozoic era. The most affected organisms were largely the microscopic organisms that float in the water column.

The structure of these micro-organisms, and the chemical composition of their shells, provides the best record of the environment at the time, and is studied extensively to reconstruct ancient climates.

On land, the PETM is associated with a distinct turnover of plant species, more so than any effects seen in the animal world. The forests, especially those at northern latitudes, are composed of uniquely different tree species before and after the event.

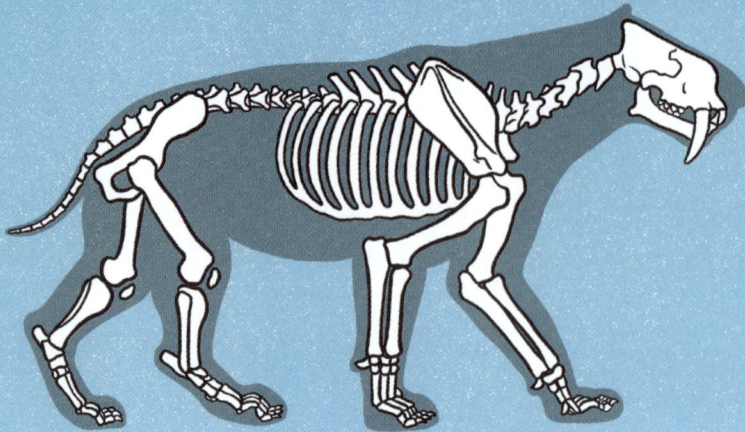

GREAT AMERICAN BIOTIC INTERCHANGE

Having been separated for 120 million years, just 20 million years ago the continents of North and South America began to collide, culminating in the formation of what we know today as the isthmus of Panama, around 3 million years ago. Once connected, the formerly isolated ecosystems of the two continents were able to merge and interact for the first time, in an ecological landmark known as the Great American Biotic Interchange.

It was the species from the north that proved the more able to take advantage of this opportunity, frequently outcompeting their challengers to the south and driving them to extinction. Unique predators of the south such as 6-foot-tall (1.8 m) flightless birds, whose large hooked beaks and long powerful legs earned them the name "Terror Birds," were still outcompeted by monstrous animals from the north like the lethal saber-toothed cats.

Herbivorous grazing mammals of the south didn't fare much better, as their northern counterparts swiftly moved into their habitats and outcompeted them for food.

Interestingly, as a result of human developments that have caused habitats to fragment and made it difficult for any large animals to cross the isthmus of Panama today, for many groups of animals, the two continents are functionally near-isolated once more.

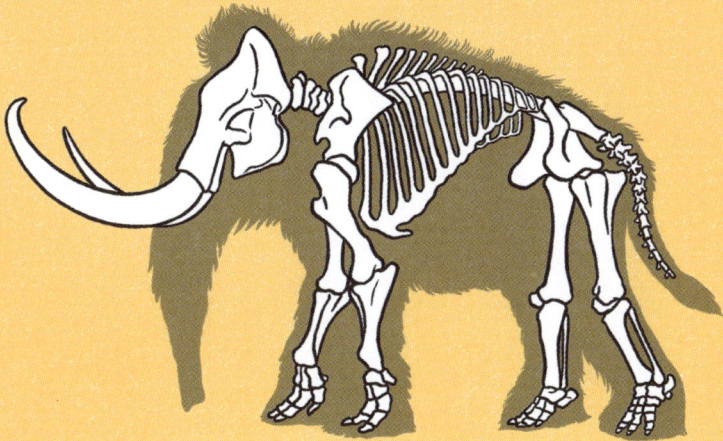

THE
ICE AGE

There have been many ice ages across the vast history of the Earth—periods when cold temperatures and glaciation extended much further from the poles than is seen today. However, the term is most commonly used to describe the most recent time this occurred on the planet, between about 26,000 and 20,000 years ago, also referred to as the last glacial maximum.

At this time, during an epoch known as the Pleistocene, seasonal ice extended from the north pole far enough south to encompass all of Canada in North America, and all of Scandinavia in Europe (including the Baltics and half of the UK and Ireland). Global temperatures were up to 10.2°F (6°C) colder than today.

Even though it is relatively so near to the modern day, many of the animals of the time were wildly different.

The grasslands of Eurasia were dominated by the giant woolly mammoths, rhinos, and elk with antlers spanning over 10 feet (3 m) across. Among these spectacular beasts also lived the early humans, those commonly referred to as "cavepeople," painting rock art and giving a fascinating insight into the evolution of culture.

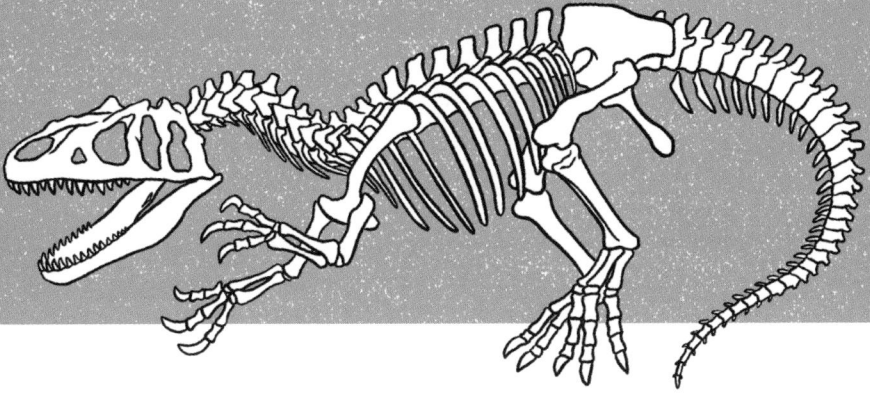

CHAPTER 4
PREHISTORIC PLANET

THE PREHISTORIC PLANET

The Earth is a dynamic planet with a constantly changing surface, meaning the globe we think of today is not the same as the one the dinosaurs would have known. In fact, being around for so many millions of years, this is true even within the dinosaurs as a group: those that first evolved in the Triassic would not have recognized the world by the end of the Cretaceous.

This chapter will focus in part on the geography of the Mesozoic era, introducing the names of the major landmasses and oceans present when the dinosaurs first evolved, and how these landmasses shifted during their reign. It will also highlight the truly massive geological features that today provide evidence as to how the era came to an end in one of the most famous catastrophes life has ever endured.

Today, dinosaur fossils can be found on every continent, and many countries host prized fossil localities, some of which will be explored here.

Despite the vast richness of the fossil record, we know there are serious biases that cannot be ignored regarding the sources of our information.

Most of these come from the odds of preservation in the first instance, where sturdier material is much more likely to fossilize, leaving many of the fragile, hollow-boned dinosaurs to crumble and decay, leaving no remains. The environment of the time plays a large role. It is much easier for animals living in a swamp to fossilize than those in the deserts, as the bodies are far more likely to be buried correctly and remain undisturbed.

Even if fossils are preserved, some collection biases are unavoidable—for instance, places where the current geography limits the accessibility of the fossils. There are certainly fossils hidden under the vast ice sheets of Antarctica, and beneath the sand dunes of the mid-Sahara, which cannot be feasibly reached, not to mention those that are residing deep under water.

In other areas, the bias comes from ourselves. Historically richer nations have many more years of study and time for exploration, as do places which place dinosaurs higher in their cultural importance. By some counts, over two-thirds of all recorded dinosaur discoveries originate from just eight countries, with the top two being, unsurprisingly, the USA and China. Countries where geopolitical instability and poverty plague the populations are largely unknown in paleontology as, quite rightly, issues of humanity and safety must take precedence.

As such, it is always worth remembering that what we see of dinosaur fossils, and have reconstructed of their age, can only ever give us the smallest snapshot of the reality. When considering all this, just how much we have been able to bring to life of the dinosaurs through research seems a true miracle.

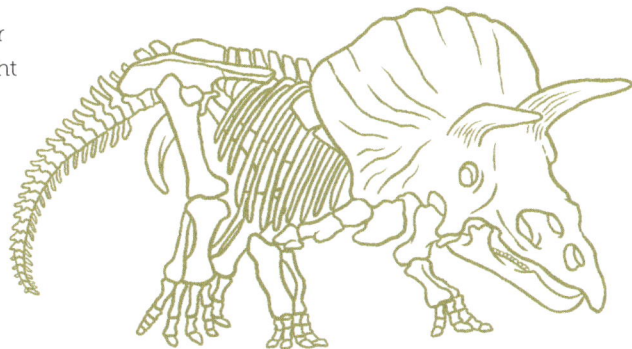

PANGAEA

At the start of the Mesozoic era, what we now think of as seven distinct continents were joined in one large landmass, known as Pangaea (meaning "All Earth"). The supercontinent began to break apart a little over 200 million years ago, with the land slowly moving closer to the positions we know them to be in today.

The landmasses we see as continents sit on units called tectonic plates. These plates are able to move due to forces deep in the Earth, as the partially molten rock that makes up the Earth's mantle rises and falls in convection currents, pulling and pushing the over-riding plates with it.

These plates are constantly moving, though at rates so slow that they are only truly visible when viewed over many millions of years.

It was on Pangaea that the dinosaurs first evolved and diversified, spreading to all corners of the landmass before it broke up completely. This is why today we are able to find dinosaur fossils on all seven continents.

The ubiquity of fossils of certain animals and plants are in fact what formed some of the earliest evidence for the existence of Pangaea—creatures which would not have been able to span the vast oceans of today having been found spanning continents. Geographical features like mountain ranges can also be traced across the continents, despite in some cases being divided by oceans.

A common misconception is that this was the original form of Earth's landmasses when the oceans first formed, but this is incorrect. Prior to being connected as one, the continents had various arrangements, including at one stage being part of a different supercontinent called Rodinia, which was not quite as cohesive as Pangaea, and broke apart in the Cambrian period.

Pangaea itself formed around 300 million years ago, and held largely together for about 100 million years. This was the most recent time that all the continents were united in this way. By modeling the most likely direction the continents will travel in the future, it is predicted that they will unite again about 250 million years from now.

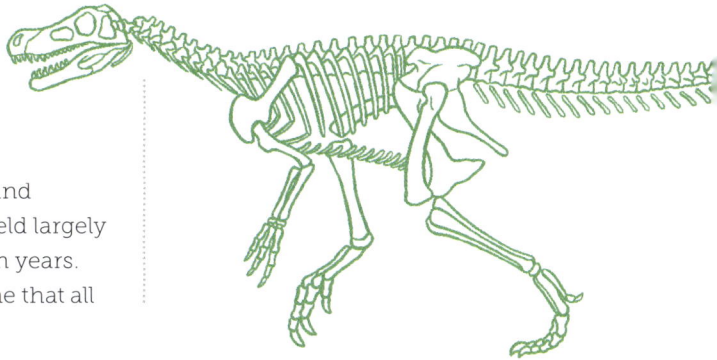

DINOSAUR ORIGINS

Pinpointing exactly where on Pangaea the first dinosaurs emerged is difficult, given that it is impossible to identify with absolute certainty the oldest dinosaur. What evidence we do have points toward the south of the supercontinent. Some of those earliest fossils include *Nyasasaurus* of Tanzania, as well as *Eoraptor* and *Herrerasaurus* of Argentina.

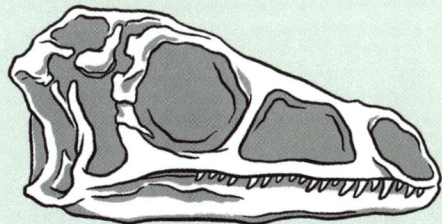

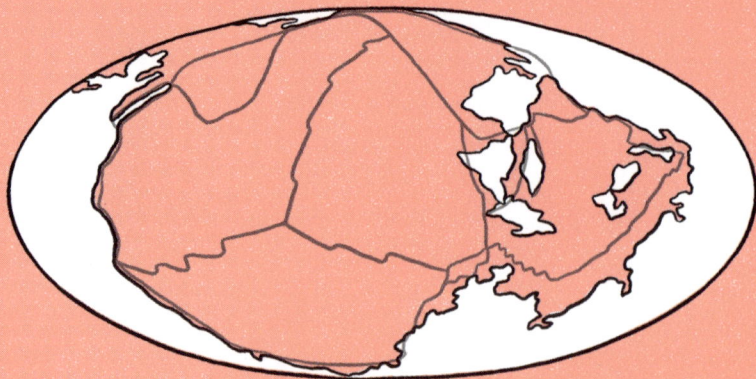

PANTHALASSA

· ·

The largest sea that life on planet Earth has ever known is the Panthalassa, fittingly named from the Greek term for "All Sea." A natural consequence of having all of the continents joined together, this body of water occupied almost 70 percent of the surface of the planet—over twice the coverage of the Pacific Ocean today (around 32 percent).

At the beginning of the dinosaur age, the Panthalassa was still at its largest. However, as the continents began to break apart and drift away from each other, it was reduced in size.

The ever-shifting nature of the tectonic plates that make up the Earth mean that even the underlying geology of this ocean has now been lost. Originally made up of three plates (the Farallon, Izanagi, and Phoenix plates), at the point where these met, a new plate was growing: the Pacific, which would go on to form the basis of much of today's Pacific Ocean.

There is much that is not understood about this vast ocean, as most of the underlying rock that formed the oceanic plates that created it has been destroyed by the natural process of plate subduction (when dense oceanic plates are dragged into the mantle underneath the continental plates).

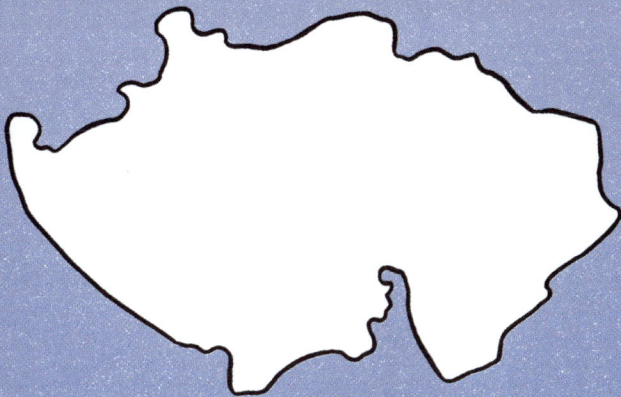

LAURASIA

When Pangaea broke up, it split into two giant supercontinents: one to the north, and one to the south. The continent to the north, made up of what is now North America, Europe, and most of Asia, we call Laurasia.

Laurasia remained fairly intact as a unit for most of the Mesozoic as, although the Atlantic was growing and would eventually pry them apart, Europe and North America stayed at least partly connected to the far north beyond the K–T extinction. The two masses only split for good 60 million years ago.

Higher sea levels in the Mesozoic meant that much of Laurasia had a very different look. A large amount of Europe was made up of many smaller islands as opposed to being a predominantly singular land mass.

North America was divided through the middle by a vast inland sea, known as the Western Interior Seaway. This split the landmass into two distinct regions: to the west was the long and narrow Laramidia; and to the east, the stouter Appalachia.

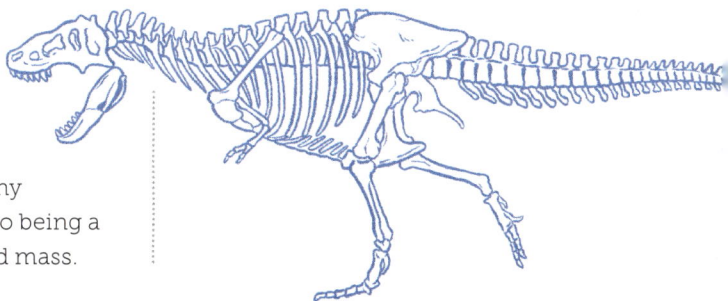

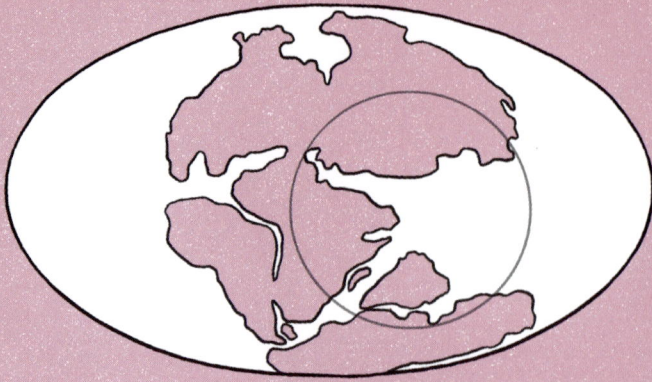

TETHYS OCEAN

One of the most important geographical features of the Mesozoic era was the Tethys Ocean. As the supercontinent of Pangaea broke up, the Tethys Ocean spanned between the northern and southern landmasses to the east, in the opening of the vaguely "C"-shaped landmass.

Through the Mesozoic, the size and structure of this ocean shifted with the continents. Notably, it was across this ocean that the Indian subcontinent traveled after breaking from the south on its voyage to what would become Asia.

Due to the difficulty of assigning precise boundaries to bodies of water, it's hard to say precisely when the Tethys Ocean ceased to be. It could be claimed that it closed when India collided with Eurasia 20 million years ago, or when the inland sea portions dried just 5 million years ago. Today, all that is left of this once massive ocean can be found in the waters of the Caspian and Black Seas.

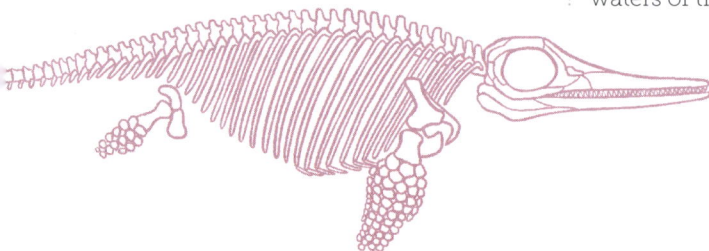

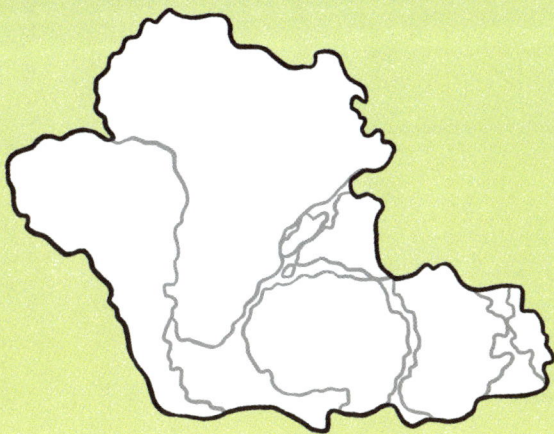

GONDWANA

While Laurasia was to the north, the southern supercontinent, made up of what is now South America, Africa, Australasia, Antarctica, and India, we call Gondwana.

Gondwana began to split apart in the Early Jurassic, when its constituent parts moved toward their current positions. One of the most impressive landmass migrations was that of the Indian subcontinent, which separated fully from the Gondwana cluster and traversed north, closing the Tethys Ocean. However, it would not collide with Asia—forming the Himalayas in the process—until well after the end of the Mesozoic era.

Likewise, Africa and Arabia would eventually drift north and collide with the Eurasian plate, though long after the Cretaceous period (becoming joined about 30 million years ago).

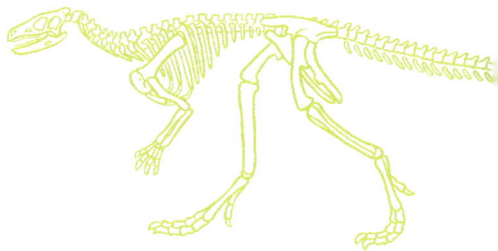

Antarctica was not at all like today either, as the lack of permanent ice caps during the Mesozoic meant that this land was not only more inhabitable for dinosaurs, but covered in forests and foliage. Unfortunately, it's hard to tell exactly what life was like there, as so many of the potential fossils left behind are locked under the vast sheets of ice we see on the continent today.

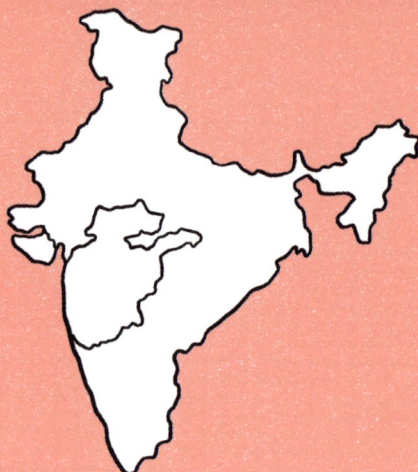

DECCAN TRAPS

An area of volcanic rock stretching 200,000 square miles (518,000 sq km) dominates the geology of Western India. These are the Deccan Traps, a natural feature that records a period of volcanic eruptions which occurred in the Late Cretaceous period. However, as the volume of rock suggests, this was far greater than any typical eruption.

The Deccan Traps are evidence of a series of eruptions across a period of approximately 700,000 years. In that time, the amount of volatile gases and other ejecta they would have produced is enormous; enough to have had a serious effect on the climate of the planet.

The timing of this event, coinciding almost exactly with the dinosaur extinction, has led many to believe this could have been one of the causes; that the dinosaurs were already struggling as a result of climate change before the asteroid hit (see K–Pg extinction, pages 130–31).

Due to the inherent bias in the fossil record, it is difficult to know for certain what diversity trends were like in the final few million years of the dinosaur age. Because of this, the scientific community remains divided on whether it truly did have a global effect. Some studies show diversity declining, whereas others suggest it remained stable.

CHICXULUB CRATER

One of the most obvious physical effects of a large asteroid striking the Earth is the enormous crater it leaves behind. For a long time, the lack of such a clearly known crater was used as evidence against the asteroid strike theory for the extinction of the dinosaurs (see K–Pg extinction, pages 130–31). It wasn't until 1991 that the crater was formally identified.

The center of the crater was located off the coast of Yucatán Peninsula and measured over 112 miles (180 km) across. It was named for a town within its radius, Chicxulub.

The crater went unnoticed for so long largely due to it being primarily off the coast, obscured by the water and 66 million years of erosion and sediment build-up. It wasn't until the development of sea-floor scanning technology that it was fully revealed.

The terrestrial section of the crater was hidden by dense vegetation and human developments, but clues to its location were there. Ringvs of caves called cenotes show where the softer rock is exposed and highlight the edge of the crater. Layers of the extra-terrestrial mineral iridium also form deposits found across the globe, which thicken the closer they are to the crater.

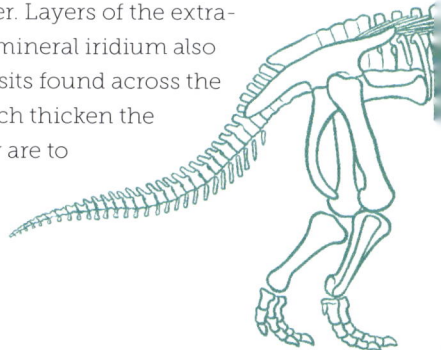

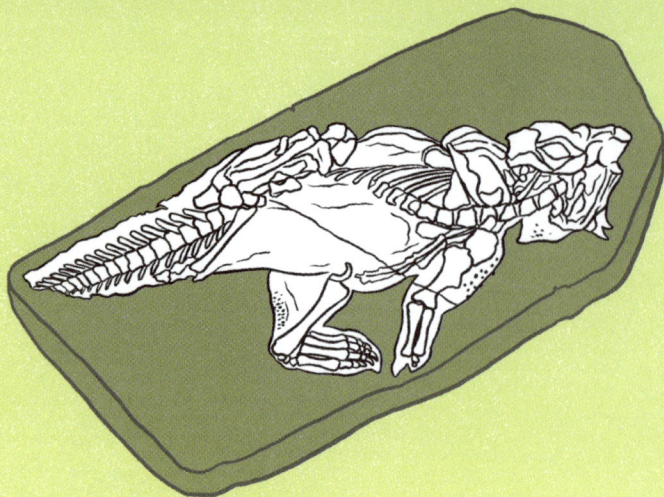

LAGERSTÄTTEN

There are many places on the planet where conditions are just right for the remains of long-dead organisms to be able to beat the odds and be preserved in superb detail. These areas are designated with the German word *Lagerstätten*: sites of exceptional preservation.

Currently there are dozens of formally identified *Lagerstätten* that can be dated to the Mesozoic era, many of them containing beautifully preserved dinosaur specimens. Unlike regular fossil sites, *Lagerstätten* may preserve features not usually seen alongside the fossil bones. Less resilient features, like skin, feathers, and even stomach contents, can be found, revealing details about how these animals truly looked and lived beyond the mere fact of their existence.

There are several conditions that can help to ensure better fossil preservation. The first is that the specimen

is buried quickly by sediment, meaning there is less time for the body to be scattered by scavengers.

It also helps if oxygen levels are low, as this restricts the spread of bacteria which decompose the soft body remains. Fossils preserved in low-oxygen sediments frequently have a dark gray or black coloration, though this is not always the case.

Once within the rock, the remains ideally shouldn't be overly disturbed. Though it may not seem like it, the rock under our feet is always moving. Stresses can fracture the rock, or even make it malleable under the right heat and pressure conditions, permanently altering the potential fossil. If too near a volcanically active area or plate boundary, the fossil could even find itself baked by intruding magma or dragged deep into the Earth and destroyed under immense pressure.

If the conditions are right for exceptional preservation, the levels of detail are stunning. Fossils can preserve even microscopic details in rock, to the point that the tiniest of wear patterns can be studied on the teeth to infer diets, and organelles invisible even under optical lenses are still visible when scanned by electron microscopes.

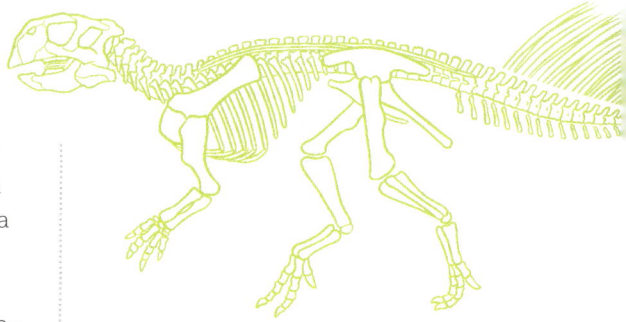

Lagerstätten have been crucial to our understanding of the dinosaurs and the world they inhabit, telling us far more than just scattered bone fragments ever could. Such a site requires enormous amounts of luck, and thankfully the dinosaurs had about 160 million years' worth of attempts to get it right.

INTRICATE STRUCTURES

Fossilized insect wings are among the most visually striking of exceptionally preserved fossils. As something we see every day, we are well aware of how fragile insect wings are.

Yet, when conditions for preservation are right, they can survive intact for millions of years in the ground, and still show the most delicate features.

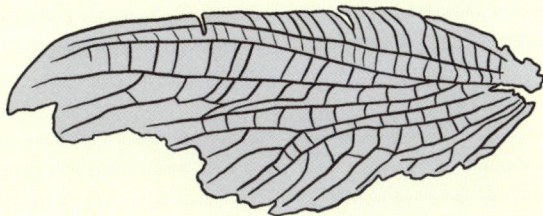

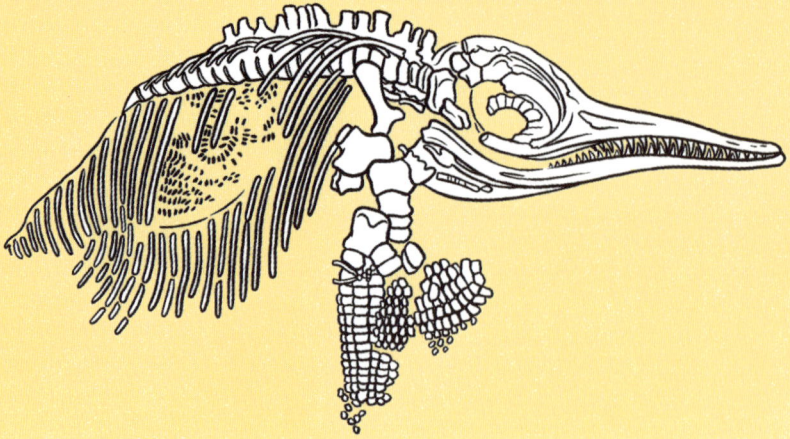

THE JURASSIC COAST

Many of the fossils instrumental in building our understanding of the ancient Earth were uncovered in a stretch of coastline in the south of England. The rocks here are so rich in fossils from the mid-Mesozoic that it has since been named the Jurassic Coast. This 95-mile-long (154 km) expanse has even been named a UNESCO World Heritage Site.

Despite the name, the cliffs of the Jurassic Coast do include sections from both the Triassic and Cretaceous, meaning every period of the Mesozoic is represented somewhere within this stretch.

Though dinosaurs can be found here, such as the sauropod *Duriatitan* and the armored *Scelidosaurus*, far more famous are the marine reptiles. Ichthyosaurs have been found here since 1811, with many of the best uncovered by Mary Anning, an English collector whose work laid the foundations for modern paleontology.

Anning's pioneering finds on the Jurassic Coast included the first known plesiosaur, the first pterosaur outside of Germany, and the first correctly identified coprolites. Amazingly, she also discovered the fossilized ink of ancient squids, which she used to sketch the very fossils she was describing.

The classic Jurassic ammonites are also found in abundance here, with one area in particular having base rock so packed with giant ammonites under foot that it is known as the "Ammonite Pavement." Another area is home to a unique fossil forest where

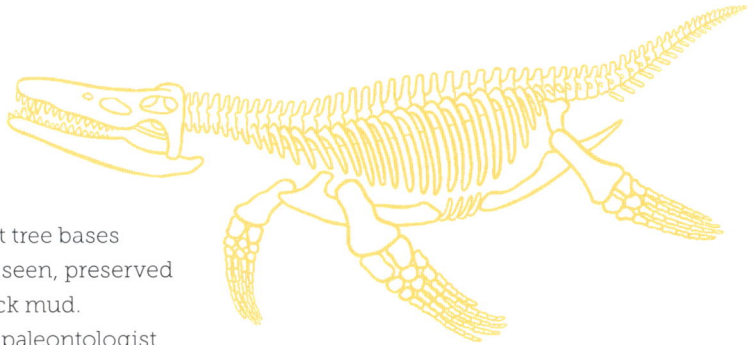

the remains of ancient tree bases of the Jurassic can be seen, preserved by being rooted in thick mud.

In 1830, the English paleontologist Henry De la Beche created the *Duria Antiquior*, depicting many of the "sea monsters" being discovered along the Jurassic Coast. This piece was the very first full scene reconstruction of a paleoenvironment, further cementing this area's reputation as being one of the most important and influential sites in the history of paleontology.

Today, the Jurassic Coast is one of the top fossil tourism hotspots in the world, and as such is frequently over-collected, with so many hands scouring the beach. And yet incredible fossils are still regularly found here. One of the more recent was a spectacular three-dimensional pliosaur skull, initially discovered in 2022.

UNIQUE INSIGHTS

When animal remains are preserved in three dimensions, as opposed to being crushed totally flat by the fossilization process, scientists have many more opportunities to understand the functionality of the animal. This enables more accurate reconstructions to be made. As such, finds like the giant pliosaur skull of Kimmeridge Bay in southern England are invaluable.

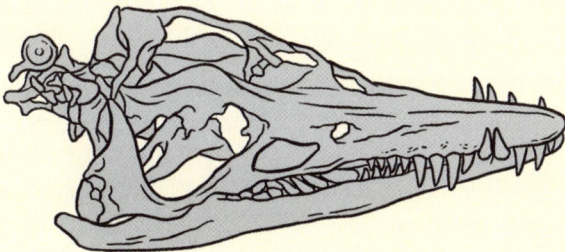

TIAOJISHAN FORMATION

One of the most spectacular assemblages of fossils from the Jurassic period can be found in the Tiaojishan Formation of China, in Hebei and Liaoning provinces. The rock preserves a subtropical forest environment in a highly volcanic region, as shown by repeating layers of ash.

The dinosaurs here lived for the most part in peace, but their habitat was constantly under threat from violent eruptions and pyroclastic flows of boiling hot ash. Many were extremely bird-like, as is shown in the spectacular fossilization of their feathered bodies. Several of them, most notably *Anchiornis*, were preserved well enough to reconstruct their color from microscopic structures in the feathers.

Clearly an evolutionary hotbed for fliers, it was also in these forests that the bizarre winged theropod *Yi* (see page 81) could be found.

Traveling across the air from tree to tree in search of small prey was a common lifestyle here, as it is not seen only in the dinosaurs. The early mammals and their relatives got in on this niche as well, as it is in the Tiaojishan Formation that the earliest gliding mammals can be found, appearing not too dissimilar from a modern flying squirrel.

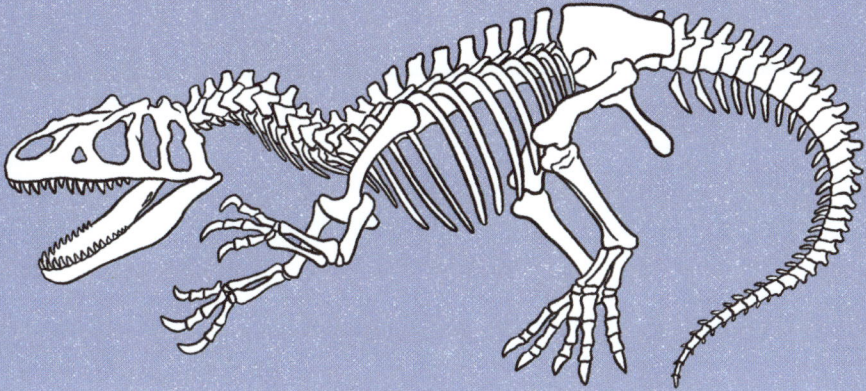

CLEVELAND-LLOYD DINOSAUR QUARRY

Declared a National Landmark of the USA in 1965, Cleveland-Lloyd Dinosaur Quarry is a vast site in Utah that has provided an enormous collection of Jurassic fossils (dated to about 152 million years ago).

Astonishingly, two-thirds of the dinosaur fossils found at this site come from the genus *Allosaurus* (see page 18). This is a far greater proportion of predators than you would ever expect to find in a healthy ecosystem.

The reason for this overabundance of allosaurs lies in the geology, which shows that the area was a frequently flooded muddy plain. Large herbivores like *Stegosaurus* may have become trapped within the mud, attracting the attention of opportunistic scavengers, who then got stuck themselves. With more and more carcasses to draw them in, a deadly feedback loop would have ensued, likely leading to the fossil balance seen today.

As for why there are so many allosaurs even compared to other local carnivores (such as *Ceratosaurus*), this could be a result of pack behavior in allosaurs, or even that the other smaller predators remained fearful of being attacked by a mostly trapped *Allosaurus*.

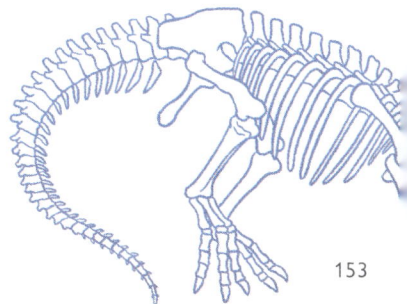

SOLNHOFEN LIMESTONE

Located in the south of Germany, Solnhofen has been famed for its Mesozoic fossils since the science of paleontology began. The rock here, initially mined for construction, preserves a Late Jurassic lagoon ecosystem.

Mostly isolated from the sea, the lagoons of this ecosystem were low in oxygen and high in salt, making them bad for decomposers and great for preserving fossils.

Many of the best fossils found here are those of invertebrates, including shrimp with every millimeter of long antennae preserved, and dragonflies complete with their delicate wings. Even creatures such as jellyfish, which usually are totally lost—their soft bodies making them almost impossible to preserve—can be found in Solnhofen.

The most famous specimen by far, however, is that of the earliest known bird, *Archaeopteryx* (see page 90), which was first described from a single feather found in Solnhofen in 1860. Other specimens associated with skeletons were soon discovered from other German quarries and remain among the most iconic and scientifically important fossils ever discovered.

The seventh discovery reported as *Archaeopteryx* was even named the "Solnhofen Specimen," although it has since been debated whether this specimen is really *Archaeopteryx* or a separate, but similar, genus.

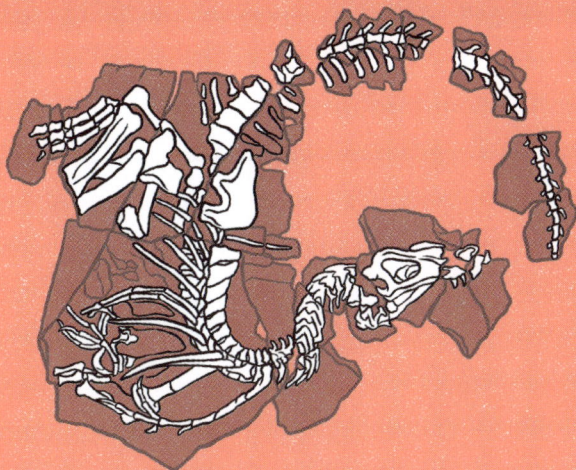

LAS HOYAS

One of the best places to find dinosaur fossils in Europe is in the Las Hoyas limestones of Spain. The rock here preserves a lake environment from the Early Cretaceous (about 125 million years ago), where many dinosaurs would live and die, their remains fossilized in the soft sediments under the water.

Las Hoyas is best known for its theropods, including the bizarre hump-backed *Concavenator*, which is now somewhat of an emblem for the location. The herbivorous iguanodon *Mantellisaurus* is also known here.

Many bird species have been discovered at Las Hoyas, though none are from the modern group we know of today. All fossil birds of Las Hoyas are *Enantiornithes* (see page 94), showing just how successful this group of "alternative birds" was during the Cretaceous period.

Much like animals today, dinosaurs living at Las Hoyas lake may have needed to be wary near the water's edge, as the preserved remains of several species of crocodilians have also been recovered here (although most were unlikely to have troubled anything other than fish, which are the most common vertebrate fossils at Las Hoyas).

YIXIAN FORMATION

Just as the Tiaojishan Formation (see page 152) is the go-to fossil location for feathered dinosaurs of the Jurassic, the Yixian Formation is the place to find feathered dinosaurs of the Cretaceous. This formation, also in China, preserves a window into a forested habitat that existed about 124 million years ago.

Though known for fossil fish for almost a century, the breakthrough that put the Yixian Formation on the global stage came in 1996 with the announcement of *Sinosauropteryx* (see page 63), the first dinosaur to have been discovered with fossilized feathers.

Since then, many more fossils of feathered theropods have come from these rocks, including the remains of *Yutyrannus* (see page 82), currently the largest animal ever to have been confirmed as having feathers.

It isn't just the feathers uniquely preserved here, as one specimen of *Psittacosaurus* (see page 59), almost certainly originally from the Yixian, shows other soft tissues preserved too, allowing for reconstruction of the skin and flesh of the dinosaur in incredible detail.

The Yixian was rich in dinosaur life, containing many ornithischians as well as a collection of sauropod bones, many of which have still not yet been fully described. Together they paint a picture of how the dinosaurs were thriving in diversity during the Early Cretaceous (see page 128).

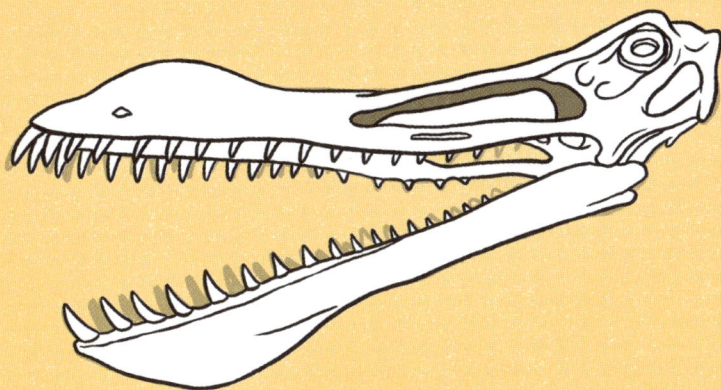

ROMUALDO FORMATION

Dated to the Early Cretaceous (110 million years ago), the Romualdo Formation of northeastern Brazil preserves a unique ecosystem at a time when South America was breaking away from Africa, opening the Atlantic Ocean.

The fossils here are preserved within nodules of limestone, formed as the oozes on the sea floor enveloped the remains of dead animals, protecting them from many of the organisms which may otherwise have disturbed their fossilization.

There are many species of dinosaur preserved here, including the spinosaurid *Irritator*, which is possibly the best known from the location.

What makes the Romualdo Formation truly special, though, are not the dinosaurs, but the pterosaurs. Over twenty species of flying reptile have been described from this formation, and their prevalence is remarkable considering how rarely pterosaur fossils are preserved, having such lightweight and fragile bones.

In recent years, fossils from this formation have been featured regularly in the news for unfortunate reasons. Frequently the target of illegal international trade, many housed in the National Museum of Brazil were also lost to a fire in 2018.

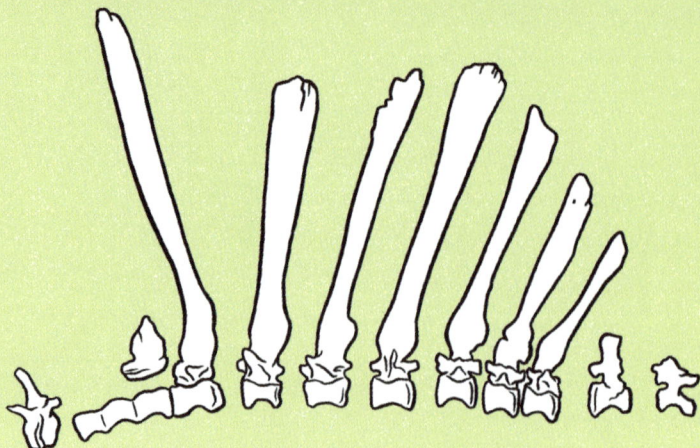

KEM KEM
GROUP

This fossil site in Morocco is known globally for producing spectacular dinosaur remains. Now a desert, the rocks preserve a frequently flooded coastal ecosystem from the Cretaceous period, around 90–100 million years ago.

Most famed of the Kem Kem dinosaurs is *Spinosaurus* (see page 65), the largest known carnivorous dinosaur. Even in this exceptional site, body fossils of this remarkable creature remain elusive, hampering reconstruction efforts. However, teeth of spinosaurs are extremely common in the Kem Kem Group, thanks to their durability and the regularity with which they were shed by the hunters.

The Kem Kem Group is an excellent hunting ground for giant predators, as several species of huge carcharo-dontosaurids can be found here.

There are plenty of teeth of smaller dinosaurs as well, showing the area was not wanting for carnivores.

The remains of herbivorous dinosaurs have been discovered at the Kem Kem site, though many of the remains are too fragmented to be identified with much confidence, beyond saying that there were certainly representatives of the Sauropoda and Ornithischia here.

Fossils from the pterosaurs in the skies above are found here in surprising abundance too, featuring some frag-ments of the truly giant azhdarchids.

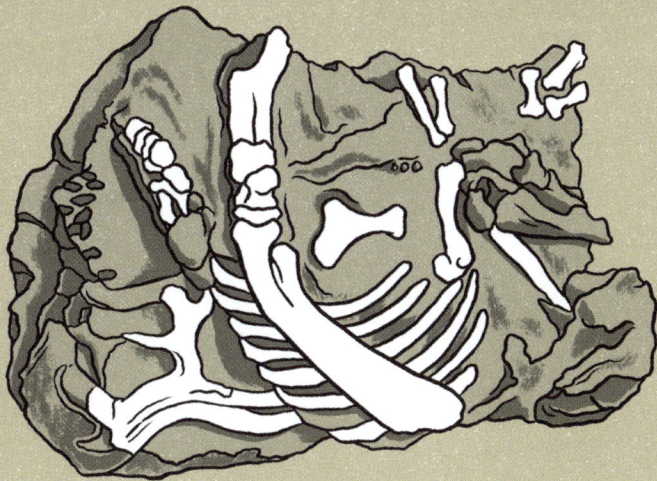

DINOSAUR PROVINCIAL PARK

Named perfectly for what you can expect to find there, Dinosaur Provincial Park of Alberta is certainly Canada's most famous fossil-hunting site. The 28-square-mile (73 sq km) area is deemed so important for our understanding of Late Cretaceous ecosystems that it is one of the rare fossil localities to have been designated a UNESCO World Heritage Site.

Many of the recognizable dinosaurs found here are large herbivores, which would have roamed the area in herds back when it was an extensive floodplain near the coast of the American inland sea, in what is now North America. They include many of the crested hadrosaurs, like *Parasaurolophus* (see page 56) and *Lambeosaurus* (see page 44).

Several species of horned ceratopsians have been uncovered at Dinosaur Provincial Park, such as the spike-frilled *Styracosaurus* (see page 68). Armored ankylosaurs are also represented, such as *Edmontia*, which was named for the region. Such dinosaurs were well-equipped to defend themselves against the large tyrannosaurids that were the top predators.

Beyond just dinosaurs, the fossils of the park capture a snapshot of a complete ecosystem, including the plant life. As a result, we know the area was rich in ferns and cycads, which provided food for the dinosaurs and shelter for the various small mammals whose teeth have been uncovered here.

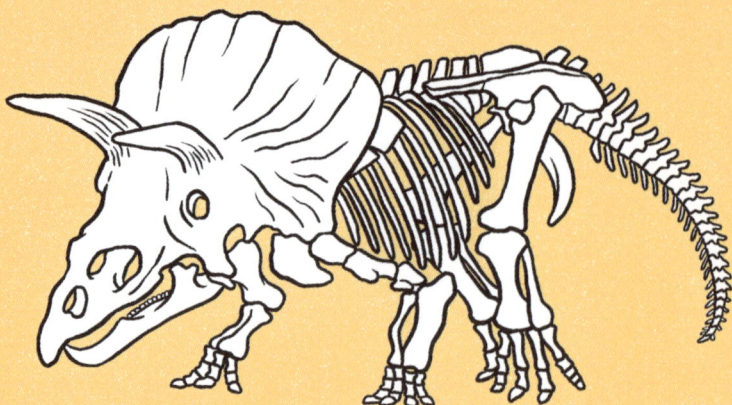

HELL CREEK FORMATION

The badlands of the Northern United States are some of the most extensively studied fossil localities in paleontology. This is largely thanks to the most famous dinosaur of all time, *Tyrannosaurus* (see pages 74–5), the fossils of which can be found in abundance here. The formation of rocks where these fossils are excavated are named for a settlement in Montana, Hell Creek, and preserve an ecosystem that existed at the very end of the Mesozoic.

We can see from the rock formations that this was once a subtropical land, with marshy and regularly flooded soils along the banks of winding rivers. It was these water systems and muds that helped preserve the fossils here so well—and not only dinosaurs, but a huge abundance of fish, plants, and turtles too.

The most common fossils found here are from ceratopsians, including *Triceratops* (see pages 72–3). Hadrosaurs are also regularly excavated here, with the most common being *Edmontosaurus* (see page 33), though *Parasaurolophus* (see page 56) is present too. Other big names of the formation include the hard-hitting *Pachycephalosaurus* (see pages 54–5) and *Ankylosaurus* (see pages 20–1).

With so many iconic dinosaur species in one location, it is fair to say that the ecosystem of the Hell Creek Formation is the closest approximation to the image instinctively conjured up when speaking about the time of the dinosaurs.

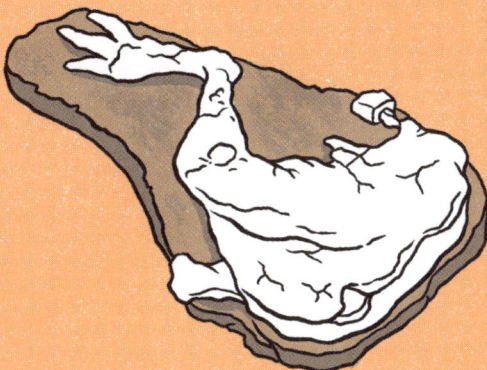

TANIS

This fossil site in North Dakota, USA, preserves a dinosaur ecosystem that existed at the end of the Cretaceous period, 66 million years ago. Yet, more than that, the remains found here are thought to be from the very final hours of the Mesozoic—literally the day the asteroid struck.

The evidence for this remarkable discovery can be found in the gills of the fish discovered there. Within them are glassy spherules that would have been ejected from the asteroid when it hit. These were gulped up by the fish and fossilized with them. Scans of the bone cross-sections can even tell us that the impact likely took place in the late spring season.

Along with the fish, there are dinosaurs at Tanis, too—notably an extremely well-preserved leg from a *Thescelosaurus*, killed by the surges that followed from the asteroid impact.

Tanis is one of the most controversial fossil sites on Earth. Doubts have been raised over the legitimacy of the geology of the area, and a huge amount of human drama surrounds the discoveries, including claims of misconduct and even falsified data in some of the analyses. Only further exploration of the site and those like it, and more paleontological finds, will help to reveal the truth.

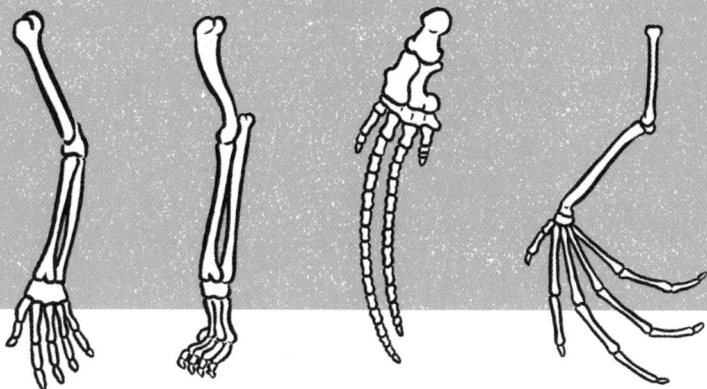

CHAPTER 5

SCIENTIFIC TERMS

SCIENTIFIC TERMS

The world of science can often appear as an incomprehensible mess of complex words and terminology which can act as a barrier to many from engaging in it. In this chapter a selection of those terms are broken down to gain a better understanding of how the modern science of paleontology works, proving that dinosaur research comprises much more than simply digging up bones and assembling them for a museum display.

Some of these terms lie at the foundations of the science—principles that have been known for hundreds of years and underpin how researchers explain the natural world. Others are relatively new, having only recently even been made possible thanks to modern technological advancements.

Among these words are some common threads. For example, several of them end with the suffix "ology," which is derived from the ancient Greek word meaning "study," and is applied to various fields of science (such as biology, the study of life).

Paleontology simply means the study of ancient life, and so applies not only to dinosaurs but to all extinct animals and plants. It isn't to be confused with archeology, which is the study of human civilization and does not have anything to do with

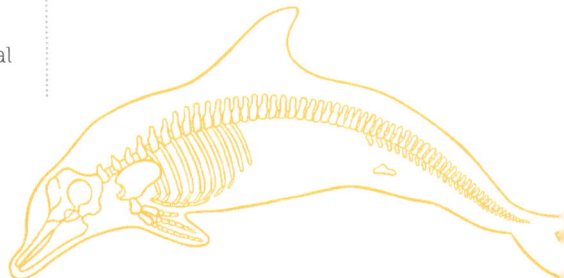

dinosaurs, despite how often the two terms are conflated in media.

Although this book focuses on dinosaurs and their time, many of the terms here are multipurpose across many fields. Paleontology is a dynamic discipline, which utilizes techniques and practices from biology, geology, chemistry, physics, and even engineering. It is only by combining knowledge from the spectrum of science that we can put together a picture of our ancient world.

Every day new discoveries are being made in the field of paleontology, and the techniques used to analyze and interpret them in the labs are getting ever more sophisticated. In the past, if a scientist wanted to see inside a dinosaur bone, they would have to smash it open and have a look. Now, scanners can allow a scientist to see internal structures without having to damage the bones at all (much to the relief of museum curators).

Understanding many of these terms can open up the world of dinosaur research beyond the big, headline-grabbing discoveries. It is often said that we are currently living through a "Golden Age" of paleontology, and that is true, but defining it as an age in this way implies it is in some way limited. The truth is that the field is gaining knowledge at an exponential rate, and it shows no signs of slowing down.

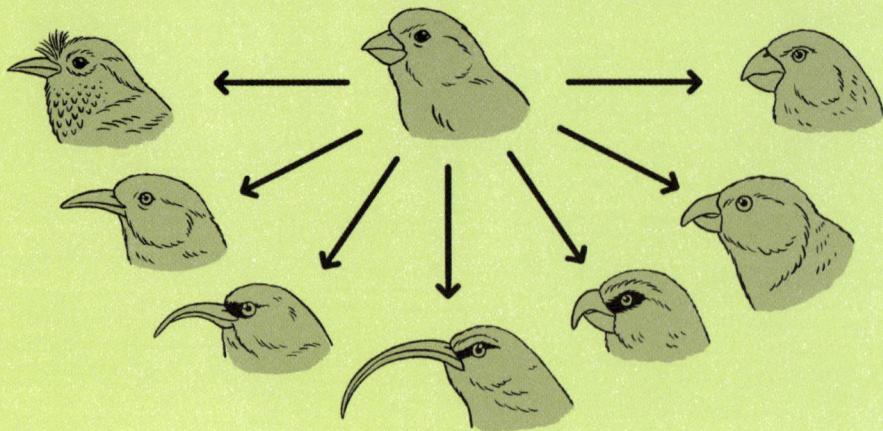

ADAPTIVE RADIATION

Every environment is full of opportunities that animals can take advantage of to survive. Each lifestyle (or niche) will require specialized adaptations in order to get the most out of it. When a single original group evolves and diversifies rapidly to fill a wide array of niches, this is known as an adaptive radiation. The number of species are literally radiating out in time from a single point.

Across evolutionary history we see many adaptive radiations, with the textbook example being of Darwin's finches. These Galapagos birds originated from one ancestor and evolved into multiple species, each with a beak perfectly adapted to best harvest their food of choice, be it short and tough to crack seeds, or long and hooked to snatch up grubs.

In paleontology, a key trigger for an adaptive radiation is the aftermath of a mass extinction, when new niches are made available due to their old occupants dying away. We see it in dinosaurs as they exploited the loss of other archosaurs after the T–J extinction, and crucially we see it in mammals, moving to replace the dinosaurs in the years of recovery after the asteroid strike (see pages 130–31).

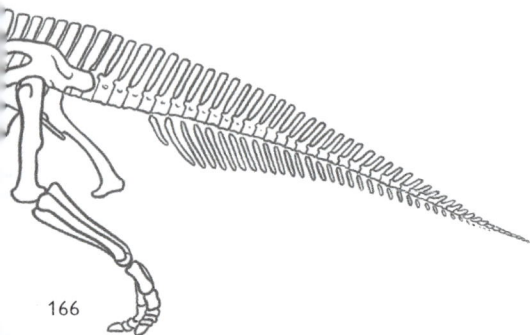

BIODIVERSITY

When studying an ecosystem, ancient or modern, scientists need a way to assess how healthy and active it is. One of the most commonly used metrics is biodiversity, often recorded as simply the number of unique species that may be found in one area.

Biodiversity is typically a measure of species richness, not a measure of how many animals there are. For example, a small pond which is home to twenty individual fish, each a different species, would be more biodiverse than a square mile inhabited by a thousand zebra and no other species.

Measuring biodiversity over time through the fossil record can show patterns in evolution, and sudden large losses in biodiversity indicate where mass extinctions have taken place. This ties biodiversity closely to the study of macroevolution.

While most uses of the word biodiversity refer to this species count, there are many different levels it can be applied to. For example, there can be diversity within the genetics of one single species, which can eventually lead to speciation events—where the differences mount to the point that individuals connected by their deep family tree become distinct species.

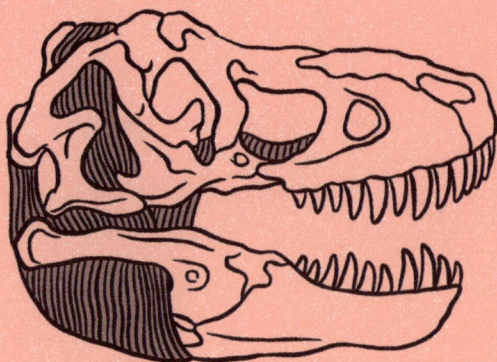

BIOMECHANICS

Integral to understanding how extinct animals lived is the knowledge of how they moved and ate, which is the science of biomechanics. Looking at the microscopic details of the bones can reveal the sites where muscles attached, allowing for accurate reconstructions of the flesh on the body, crucial for understanding how the creature may have walked, for example.

This field borrows techniques from engineering and physics, such as using computer modeling to test the maximum stresses that bones can withstand. Many of the same principles that are used to test the load-bearing capabilities of bridges can be applied to testing the stress resistance of bones. Practical experiments can be conducted as well, such as puncturing ballistic gels with replica teeth to examine hypotheses about how they may have been most effectively used in feeding. Models can be put through wind tunnel experiments to test how they could fly or glide—a practice just as useful for winged dinosaurs as it is for modern planes.

Biomechanics is a quantifiable science based on thoroughly examined evidence, far from the guessing game many assume reconstructions of dinosaur movements and capabilities to be.

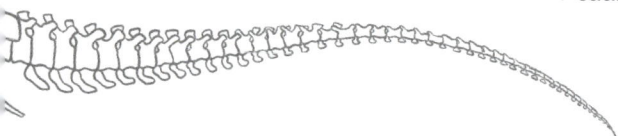

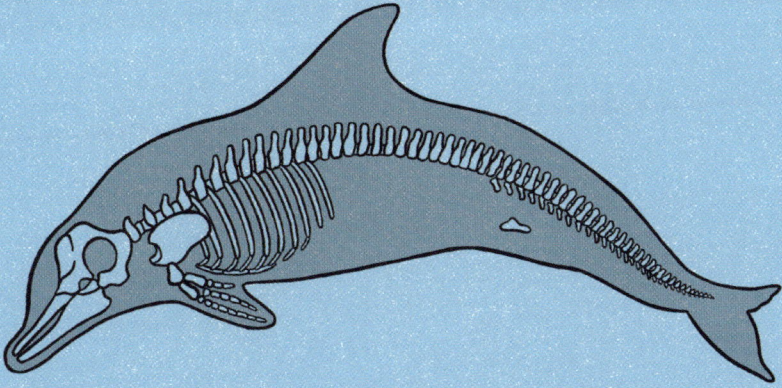

CONVERGENCE

It is common in nature to see species that share remarkably similar features, but which are not necessarily closely related. This is known as convergence (or convergent evolution), and occurs as a result of animals evolving the same adaptations when faced with the same environmental pressures.

An example is the body shape of vertebrates with aquatic lifestyles. The combination of a streamlined body shape, extended dorsal fin, and paddle caudal (tail) fin for propulsion has been adopted by many fish, reptilian ichthyosaurs, and mammalian dolphins and whales. None of these groups are closely related to one another; they all evolved these features independently of one another.

Convergence can be a problem for paleontologists. Unable to perform genetic tests on fossil bones, when superficially similar organisms are discovered, they can mistakenly be thought to share a close family relation. As such, it is important to look at as many factors as possible when describing and defining a species. Thankfully, there will always be telltale signs that give away an animal's true family origins, even if it is something easily overlooked on a casual glance, such as the shape of the wrist joints, or the presence of an extra bone at the end of the jaw.

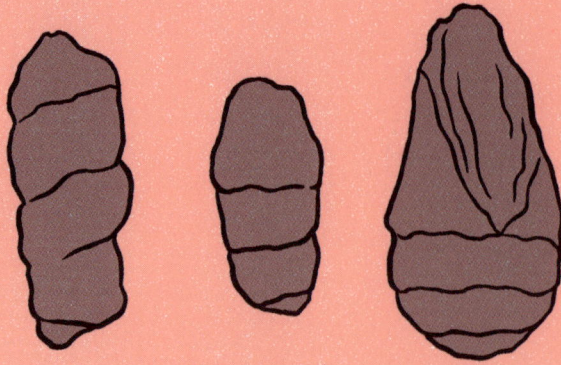

COPROLITE

When animals die, it is not only the physical remains of their bodies that might be preserved as fossils, but also the waste products they leave behind. Dinosaur dung, too, has a chance of being buried, mineralized into rock, and preserved for millions of years. The resulting fossil is known as a coprolite.

Though it was undoubtedly something to avoid during the Mesozoic, once fossilized, dung can be invaluable as a scientific tool for understanding not just the anatomy of prehistoric creatures, but also their behaviors.

The preservation detail is often such that microscopic analysis of internal coprolite structure can reveal the diet of the producer.

Sometimes features (such as the most delicate of animal bones) have a better chance of being preserved once "protected" within the coprolite than they would on their own. So, ironically, an animal may be better preserved in the future because it was eaten in the past.

The tricky part for paleontologists is matching the coprolite to the animal that produced it, as, unless it is directly associated with a body, it is near impossible to be certain.

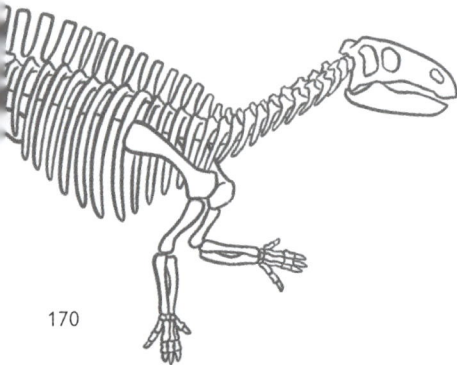

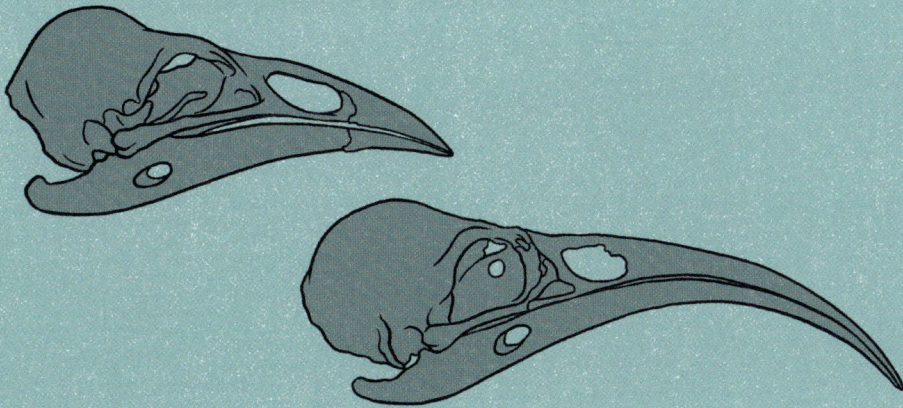

DIMORPHISM

Variation between individual animals doesn't always mean they are different species. One of the best examples of this is in domestic dogs, which come in a huge variety of different shapes and sizes, despite all being the same species, *Canis familiaris*. In science, we call such internal variation "dimorphism."

A common form of this is sexual dimorphism, when animals look different depending on their sex. Though the differences are often striking in life, such as the vibrant color of male birds of paradise compared to the females' rather drab plumage, they are much harder to detect when only fossil bones are available for study. As such, dimorphism in dinosaurs has always been hard to prove.

There are some instances where sexual dimorphism has been inferred from the diversity in limb sizes in stegosaurs, like *Kentrosaurus* (see page 43), and the crests of certain hadrosaurs; however, none of these can be definitively proven.

Sexual dimorphism is extremely common in the dinosaurs' closest living relatives, the birds, so it is logical to assume it was present in their ancestors. Several species show extreme differences between the sexes, like the extinct huia of New Zealand, whose males had extremely long, curved beaks compared to females.

GASTROLITH

Some animals require additional help beyond just their teeth to help them physically grind tough food. One solution seen across the animal kingdom is to swallow small stones that can act as grinding mills within their stomachs. These stones are known as gastroliths.

The act of mashing together with the food, combined with the stomach acids, gives them a smooth and almost polished texture. Of course, even so, it can often be hard to distinguish them from the surrounding tons of rock.

The best way to identify gastroliths is if they are found in association with the animal itself, among its bones. Another clue would be if they are of a totally different rock type to what should be found normally in the area, which would imply they have been transported.

A number of dinosaurs are known to have utilized gastroliths, including iguanodonts, and some finds have been used to infer migration behavior in sauropods.

Beyond dinosaurs, plesiosaurs are known to have used gastroliths often as well. However, instead of aiding in digestion, they swallowed these stones to act as ballast to aid them in their marine home.

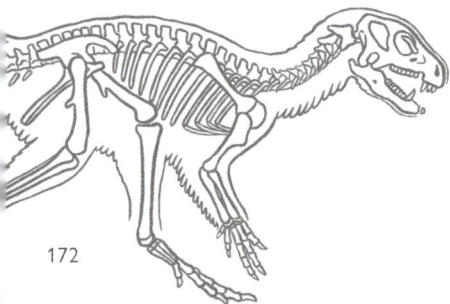

HOLOTYPE

A holotype is the individual fossil specimen upon which a new species is named. The descriptions of these special fossils define the characteristics of a species. Any new discoveries are compared directly to the holotype specimen to assess whether or not they belong to the same species.

Usually stored safely in museums, the measurements and features of the holotype are published in detail so that the information is available for anyone to check when they suspect they may have found another individual.

Holotypes are extremely valuable specimens for science, and some examples have even become celebrities in the fossil world. One of the best known is the holotype of *Megalosaurus*, a single jaw bone with multiple spiked teeth that was the first ever dinosaur fossil to be formally described and named by science.

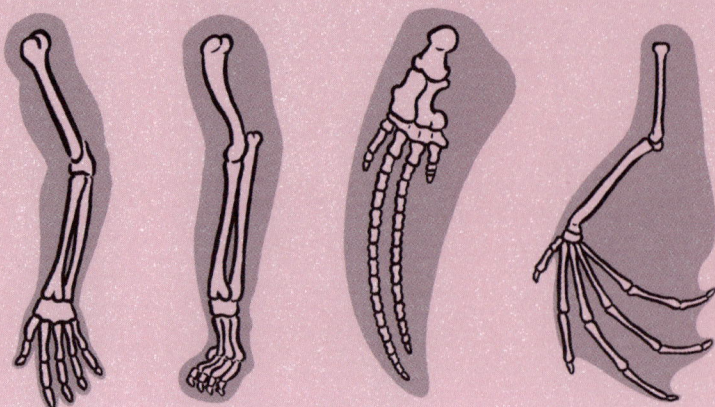

HOMOLOGY

Homology refers to structures and anatomical features common to multiple species that share a common ancestor. Due to both groups having evolved from the same body plan, the core features are still present, even if they have adapted to fulfill very different functions.

An example of homology can be found in the arms of terrestrial vertebrate animals. The arms of humans, bats, and whales are very different in size and shape, but the number and relative position of the bones remains the same. All have a humerus, followed by a radius and ulna, and the same wrist and hand bones. The bones are all present, even though each animal's appearance is starkly different. This is because the species all evolved from a single ancestor that possessed those same bones. Even birds and dinosaurs have these features.

Homology is evident across the tree of life, and can be useful for finding family connections. A notable example is the skeletal structures that birds and dinosaurs share, which indicates that they both evolved from the same ancestor.

Sometimes homology can be an evolutionary hindrance that prevents species from adapting to environments as efficiently as they might otherwise. Bone structures inherited from ancestors far back in the family tree can place physical limits on an animal's ability to adapt.

ICHNOFOSSILS

In life, an animal leaves behind traces of how it interacted with the world around it, and, if conditions are right, these can become fossils. Even something as simple as a footprint left in wet mud can be preserved. Such remains are known scientifically as ichnofossils, or, more commonly, "trace" fossils.

Ichnofossils can reveal a lot about an extinct animal's behavior. The most basic could be observing a large set of footprints together and inferring herding, whereas more complex measurements of distance between prints can be used in calculations of the dinosaur's walking speed.

Burrowing ability is another aspect of behavior that ichnofossils can reveal. It's all well and good testing to show that a dinosaur would be capable of burrowing, but finding the burrow gives solid proof of behavior. And the structure of the burrow might even provide clues about its family life, and how it raised its young.

Sometimes ichnofossils can be very easy to miss, and inferences should be made with caution. For example, alongside clear pterosaur footprints have been found sediment puncture and raking marks, which were potentially made by the pterosaurs's beaks probing the mud for invertebrates, though such theories are very difficult to prove definitively.

LAZARUS
TAXA

Because of the patchy and imperfect nature of the fossil record, it is possible for there to be large gaps of time between appearances of species. These gaps can be multiple millions of years, appearing almost as though the animal has returned from extinction. When this occurs, they are given the name Lazarus Taxa, after the biblical story of a man who was raised from the dead.

The most famous and remarkable example of Lazarus Taxa is that of the coelacanth. This order of lobe-finned fish was last known in the fossil record from the Cretaceous period, around 80 million years ago, having apparently then become extinct. This was until 1938, when a freshly caught specimen was found being sold in a market in South Africa.

Incredibly, at the same time that scientists in Europe were confidently declaring coelacanths to be an obscure animal which died out with the dinosaurs, fishermen off the Comoro Islands were regularly catching them live.

As well as being interesting stories to tell, the existence of Lazarus Taxa more than anything shows how unreliable the fossil record can be. There is always more to discover, and what is known is just a small sample of true biodiversity.

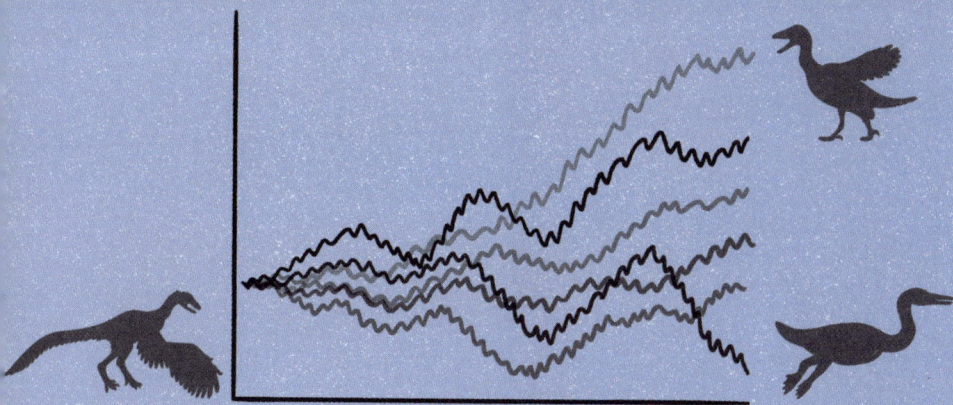

MACROEVOLUTION

Evolution is the process by which species arise over time, adapting from common ancestors to better suit their environment or improve certain functions. However, that is thinking very much at the level of a single species. Macroevolution is the study of whole patterns of evolution across multiple lineages, across different times and environments.

For example, saying that one lineage of north-dwelling brown bears evolved into polar bears is evolution, whereas saying that dinosaurs evolved into birds would be macroevolution. One involves the adaptation of a few features; the other involves funda-mental change to whole regions of the body, including the develop-ment of several novel features.

Studying evolution as a whole can reveal surprising things, like bursts in evolutionary rates in certain lineages when environmental changes occur, or a novel mutation leading to the ability to exploit many new niches.

In this way, macroevolution is very much tied to the idea of adaptive radiations as one of the recognizable patterns. Others include stability, or highly directional change, being driven by one or a combi-nation of particular external factors acting on the populations.

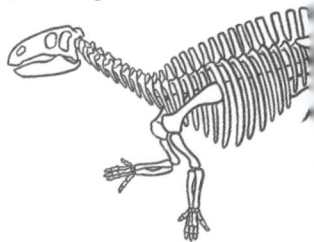

MELANOSOMES

Animals today come in a multitude of different colors, and there is no reason to think that the dinosaurs weren't just as vibrant. Until recently, the color of dinosaurs could only be guessed at, until microscopic structures called melanosomes were discovered preserved intact in the feathers of long-extinct dinosaurs.

Melanosomes are structures which house the pigment melanin, the same pigment that colors our own hair and skin. In modern bird feathers, different shapes and orientations of melanosomes give rise to different colors. Elongate melanosomes create a range of black, gray, or blonde colors depending on their concentration, whereas the spherical melanosomes correlate to reddish hues. Where there is no coloration, there are no melanosomes present.

The exact same patterns in melanin have been seen in dinosaur feathers, and have made color-accurate reconstructions of dinosaurs possible. Though initially observed some time ago, it was only in the twenty-first century that imaging technology advanced to the point that they could be confirmed as melanosomes, over previous assumptions that they were bacteria.

Unfortunately, because this reconstruction technique requires truly exceptional preservation, it is only possible in a handful of dinosaur specimens.

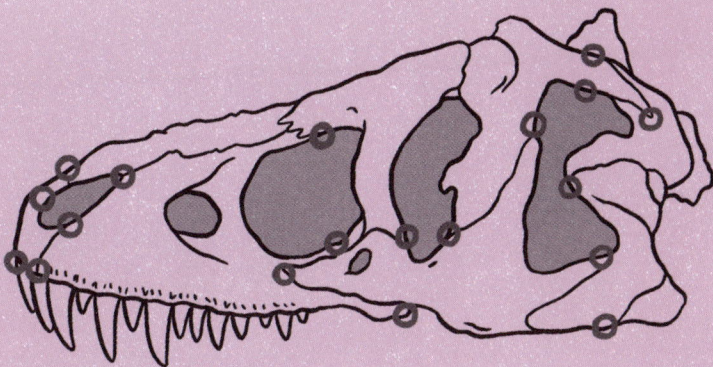

MORPHOLOGY

The scientific term for the description of shape in nature is morphology, literally meaning "the study of form." Morphology can refer to an animal's entire body shape, or its individual features.

Particular morphologies are favored by different lifestyles, enabling paleontologists to infer behavior in long-extinct animals by analyzing shapes. In the simplest of examples, when looking at teeth, it can be inferred that a dinosaur with a conical, blade-like tooth was a carnivore, whereas a flat shearing tooth would be ideal for cutting through vegetation.

As things get more complex, more intricate behaviors and interactions can be theorized about. By measuring certain fixed points of morphology, like the apex of the tooth (a technique called "landmarking"), biological shapes can be quantified and plotted like data for direct comparison between species.

When plotted, the area occupied by one group is called a "morpho-space," and can be checked against other species. If there is an overlap between morphospaces, it implies that their function was similar, and the two groups may have been in competition with one another, driving their evolution.

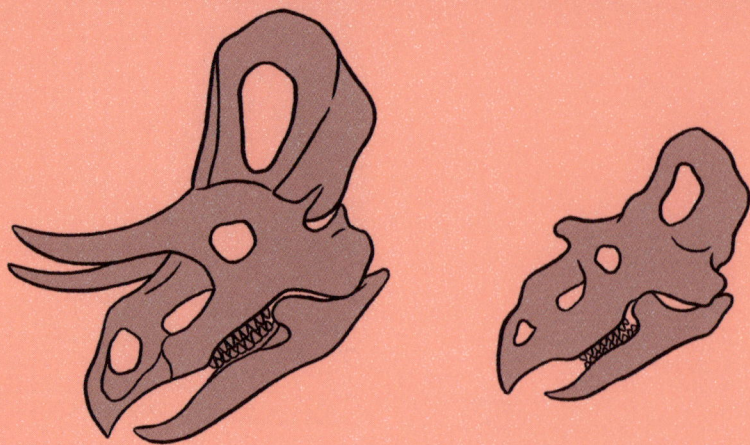

ONTOGENY

Many animals change substantially as they age, and this process of development is known as ontogeny. This commonly involves the growth of elaborate structures for display and competition, which can massively alter the animal's appearance. With only fragmented skeletal remains to interpret, however, this can be a huge problem when studying extinct animals.

One of the more famous debates in this area centers around *Triceratops* (see pages 72–3), which some have argued is merely a juvenile form of *Torosaurus* (a dinosaur which looks somewhat similar, though has large gaps in the frill—unlike *Triceratops*'s, which is solid). Though a popular idea in the media when first proposed, this theory has since been highly rebutted and seems very unlikely.

There are ways of testing for a dinosaur's age, such as looking in microscopic detail at the structure of the teeth and bones, where patterns of seasonal growth or extensive wear can be detected. This also allows us to get some estimates of dinosaur longevity, with the longest-living dinosaurs believed to be the large sauropods, who may have lived for many decades (possibly even close to a century).

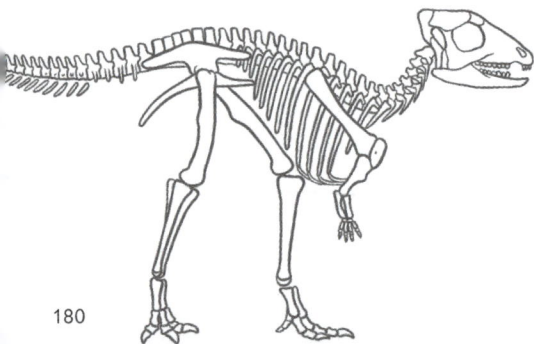

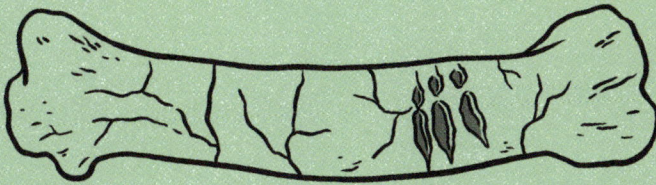

PATHOLOGY

The way an animal lives will often leave physical reminders on the body. Scaring and breakages can often damage bone, and be forever preserved in the fossil. When a scientist refers to the pathology of a dinosaur, they are referring to these remnants.

All pathologies give insight into animal behavior. This could be the bite marks left by an attacking predator on the bones of the prey, or even a simple break from a dinosaur having had an accidental fall.

Layers of bone regrowth are easily spotted, as they are not as neat as the initial bone (they tend to be more globular) and have a different structural composition. Sometimes, rather than extra bone, it is the lack of it that is interesting, as bone loss or deformation can imply that a dinosaur was suffering from a disease or infection.

Where a bone has healed can also reveal behavior, as it suggests the dinosaur must have survived whatever injury struck it long enough to heal. Famously, this is seen in some large herbivorous dinosaurs from the Late Cretaceous, discovered in North America, suggesting they were being actively hunted by *Tyrannosaurus*, from which we can surmise that it was not purely a scavenger.

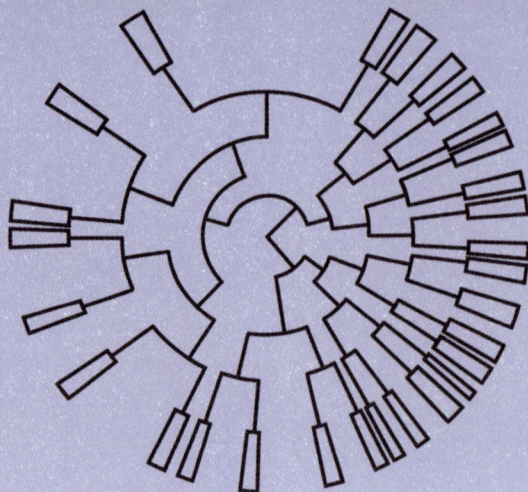

PHYLOGENY

A crucial aspect of studying extinct life is understanding how the species relate to each other and animals alive today. A family tree of evolutionary relationships is known as a phylogeny.

Phylogenies are always highly controversial in dinosaur paleontology, as there is no genetic data available for testing (in living animals, their DNA can be used to analyze family relations). Instead, a combination of factors must be used, categorizing as many individual physical attributes as possible and feeding them into computer systems which analyze them and make a tree of most likely family relationships, using the known fossil data as anchor points.

This process can involve huge amounts of data and take a very long time to compute, as possible variations increase exponentially the more animals and features are added to compare. But if the same relationships are resolved by the system time and time again, it can be said with more certainty that a true relationship has been found.

Even so, it's a very tricky and complicated process, which is why the phylogenies that paleontologists draw up for extinct animals like dinosaurs are constantly changing and being redrawn as new information comes to light, and more analyses are run.

RADIOMETRIC DATING

The determination of the ages of rocks is not a guess or a vague estimate, but a rigorously calculated scientific measurement done via the process of radiometric dating.

Many rocks contain radioactive elemental isotopes, like uranium. These isotopes are unstable and will decay to form stable varieties (in the case of uranium, it decays to lead). This process is not random, and occurs at a precise rate which can be calculated. The time it takes for half of the unstable "parent" isotopes to decay to stable "daughter" isotopes is known as its half-life.

If a half-life is known, then measuring the precise number of parent and daughter isotopes in a sample can tell you the age of a rock. This is only possible on igneous rocks, however, where new crystals have formed by volcanic processes. Volcanic layers and samples are therefore hugely important for dating fossils relatively, which are found in sedimentary rocks.

A commonly talked-about example of this is carbon dating, yet the carbon isotope involved has a half-life of just 5,730 years, far too short to be used on anything as old as Mesozoic rocks. Carbon dating is therefore mostly used for studies of ancient humans, and not dinosaurs.

STRATIGRAPHY

The rock seen in cliffs and under our feet is not a single unified mass, but split into clearly defined layers and structures. This is what scientists refer to as stratigraphy, and it helps to piece together a complete timeline of Earth's history.

Rocks are laid down in horizontal layers, one on top of another. Each unit is known as a bed, and a collection of beds make up the strata, a rock sequence. This means that, unless inverted by huge tectonic forces, rocks higher up in the sequence are younger than those lower down, allowing fossils to be dated relatively without any extra analysis.

Studying stratigraphy can reveal environmental details in an instant, too. Red-colored sandy rocks are rich in iron and indicate arid environments, whereas black silty rocks usually show a marine deposit with low oxygen. The chemical composition can further tell us about the environment in which it formed, giving more clues as to how the fossils found there lived in the past.

Here, stratigraphy is only touched on for a brief overview, but it is the gateway to the incredible world of geology, which can be used to uncover endless secrets of the past Earth.

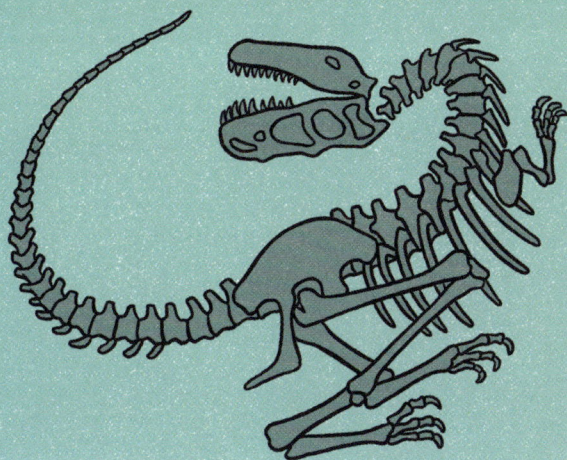

TAPHONOMY

The transition from living tissue to fossil remains is not a simple process, and there are countless factors that can influence the outcome, and whether it happens successfully at all. The study of these processes is known as taphonomy.

The first stages of taphonomy involve what happens immediately after the dinosaur dies. They may be scavenged by opportunist dinosaurs and mammals, their remains scattered around in the process.

Speed of burial in sediment, the availability of oxygen (required by decomposing microorganisms), and composition of the body all control how the dinosaur decays, in turn dictating how much might be preserved.

Once in the sediment, the organic material is slowly replaced by minerals, precipitating out of fluid into the space occupied by the body, turning it to fossil. Geological processes might even deform the shape of the body post-burial, introducing difficulties for reconstructions.

If they do remain exposed and intact, it is common to see dinosaurs adopt the "decay posture." As their muscles and ligaments contract, their head and tail are drawn in to each other, pulling over their back. Often the case in small specimens, this posture can also be seen in giants, as in the famed *Tyrannosaurus* specimen "Black Beauty."

TAXONOMY

In order to study the animals of the past, paleontologists must classify them into distinct species. This is the process of taxonomy, the science of grouping similar animals and naming them.

Animals are classified with a binomial system, two words forming its scientific name. The first word tells us its genus, the second the species. Modern creatures have these binomial names too, though they are rarely used in favor of their common names. For example, most would sooner recognize "giant anteater" than its scientific name, *Myrmecophaga tridactyla*.

Dinosaurs are unique, however, in that they are known almost exclusively by these scientific binomials. Though in almost all cases it is only the genus name, not the species, that is most commonly used. The glaring exception to this is, of course, *Tyrannosaurus rex*.

Beyond genus and species, taxonomy categorizes animals into wider groups at various other levels. The classical order (from broadest to most specific) is kingdom, phylum, class, order, family, genus, species.

This system is now antiquated (thanks to our understanding of modern genetics), though does still largely hold and is commonly used. Another term regularly used is clade, which simply refers to any defined group of animals who share a common ancestor in their evolutionary history.

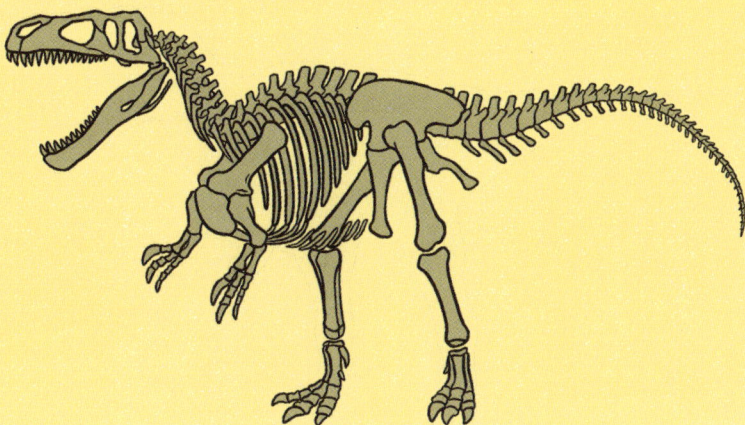

WASTEBASKET TAXA

Paleontologists are regularly tasked with identifying animals from the smallest fragments of fossils. As such, mistakes are frequently made and not always corrected. Eventually, fossil collections of a species can be so full of dubious material that any fossil pertaining to be that species must be taken with a grain of salt. Any species unfortunate enough to find itself in this situation is known as a Wastebasket Taxon—a group that it seems any old fossil has been thrown in with.

A good example of a Wastebasket Taxon of dinosaurs is Megalosaurus. Being the first dinosaur ever described formally, nobody at the time knew just how diverse the dinosaurs were. Without that wider context, when other large ancient bones were discovered, they would often be described as *Megalosaurus*, simply because nobody could fathom what else they could be.

The existence of Wastebasket Taxa means that, today, breakthroughs in paleontology can be made without the need to go out and dig up new fossils at all. Simply by re-examining bones left in museum stores for over a century, scientists might discover a whole new species in the 2020s that was actually found a hundred years earlier.

FURTHER READING

Bainbridge, David. *Paleontology: An Illustrated History.* Princeton, NJ: Princeton University Press, 2022.

Barrett, Paul M. *A History of Dinosaurs in 50 Fossils.* London: Natural History Museum, 2024.

Benton, Michael J. *The Dinosaurs Rediscovered: How a Scientific Revolution is Rewriting History.* London: Thames & Hudson, 2020.

Benton, Michael J. *Vertebrate Paleontology.* Hoboken, NJ: Wiley, 2024.

Benton, Michael J. *Dinosaur Behavior: An Illustrated Guide.* Princeton, NJ: Princeton University Press, 2023.

Benton, Michael J. *Dinosaurs: New Visions of a Lost World.* London: Thames & Hudson, 2021.

Brusatte, Stephen. *Dinosaur Paleobiology.* Hoboken, NJ: Wiley Blackwell, 2012.

Brusatte, Stephen. *The Rise and Fall of the Dinosaurs: The Untold Story of a Lost World.* Boston: Mariner Books, 2018.

Brusatte, Stephen. *The Rise and Reign of the Mammals: A New History, from the Shadow of the Dinosaurs to Us.* New York: Picador, 2022.

Charles, Rhys. *Frozen in Time: Fossils of the United Kingdom and Where to Find Them.* London: Orion, 2022.

Charles, Rhys. *The Little Book of Dinosaurs.* Princeton, NJ: Princeton University Press, 2024.

Erwin, Douglas H., and James W. Valentine. *The Cambrian Explosion: The Construction of Animal Biodiversity.* New York: Bedford/ Saint Martin's, 2013.

Fastovsky, David E., and David B. Weishampel. *Dinosaurs: A Concise Natural History.* Cambridge: Cambridge University Press, 2021.

Gee, Henry. *A (Very) Short History of Life on Earth: 4.6 Billion Years in 12 Chapters.* New York: Picador, 2021.

Halliday, Thomas. *Otherlands: A World in the Making.* London: Penguin, 2022.

Hone, David. *The Future of Dinosaurs: What We Don't Know, What We Can, and What We'll Never Know.* London: Hodder & Stoughton, 2022.

Hone, David. *The Tyrannosaur Chronicles: The Biology of the Tyrant Dinosaurs.* London: Bloomsbury Sigma, 2016.

Hone, David. *Uncovering Dinosaur Behavior: What They Did and How We Know.* Minneapolis: Highbridge, 2024.

Knoll, Andrew H. *A Brief History of Earth: Four Billion Years in Eight Chapters.* Boston: Mariner Books, 2021.

Naish, Darren. *Ancient Sea Reptiles: Plesiosaurs, Ichthyosaurs, Mosasaurs and More.* London: Natural History Museum, 2024.

Naish, Darren. *Dinopedia: A Brief Compendium of Dinosaur Lore.* Princeton, NJ: Princeton University Press, 2021.

Naish, Darren, and Paul M. Barrett. *Dinosaurs: How They Lived and Evolved.* London: Natural History Museum, 2023.

Panciroli, Elsa. *Beasts Before Us: The Untold Story of Mammal Origins and Evolution.* London: Bloomsbury Sigma, 2021.

Paul, Gregory S. *The Princeton Field Guide to Dinosaurs.* Princeton, NJ: Princeton University Press, 2024.

Paul, Gregory S. *The Princeton Field Guide to Predatory Dinosaurs.* Princeton, NJ: Princeton University Press, 2024.

Taylor, Paul D. *Fossils: An Essential Guide.* Chicago: University of Chicago Press, 2025.

Witton, Mark P. *King Tyrant: A Natural History of Tyrannosaurus rex.* Princeton, NJ: Princeton University Press, 2025.

Witton, Mark P. *Pterosaurs: Natural History, Evolution, Anatomy.* Princeton, NJ: Princeton University Press, 2013.

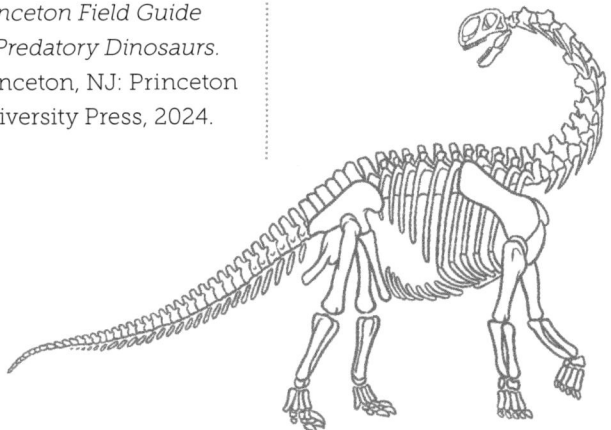

INDEX